W9-BCS-852

THE WEAPON WIZARDS

THE
WEAPON WIZARDS

How Israel Became a High-Tech Military Superpower

YAAKOV KATZ AND AMIR BOHBOT

St. Martin's Press   New York

THE WEAPON WIZARDS. Copyright © 2017 by Yaakov Katz and Amir Bohbot. All rights reserved. Printed in the United States of America. For information, address St. Martin's Press, 175 Fifth Avenue, New York, NY 10010.

Art on title page: *Israeli Air Force fighter jets fly over Masada in 2012.* IDF

www.stmartins.com

Library of Congress Cataloging-in-Publication Data
Names: Katz, Yaakov, 1979– author. | Bohbot, Amir, author.
Title: The weapon wizards : how Israel became a high-tech military superpower /
    Yaakov Katz and Amir Bohbot.
Description: New York : St. Martin's Press, [2017]
Identifiers: LCCN 2016037567| ISBN 9781250088338 (hardback) |
    ISBN 9781250088345 (e-book)
Subjects: LCSH: Israel—Military policy. | Military art and science—Israel. |
    Military weapons—Israel. | Cyberspace operations (Military
    science)—Israel. | Technology and state—Israel. | BISAC: TECHNOLOGY &
    ENGINEERING / Military Science. | HISTORY / Middle East / Israel. |
    BUSINESS & ECONOMICS / International / General.
Classification: LCC UA853.I8 K384 2017 | DDC 338.4/76234095694—dc23
LC record available at https://lccn.loc.gov/2016037567

Our books may be purchased in bulk for promotional, educational, or
business use. Please contact your local bookseller or the Macmillan Corporate and
Premium Sales Department at 1-800-221-7945, extension 5442, or by e-mail at
MacmillanSpecialMarkets@macmillan.com.

First U.S. Edition: January 2017

10   9   8   7   6   5   4   3   2

To our families,
who provide us with inspiration and hope.

CONTENTS

PREFACE

Our work on this book began with a conversation we had in the spring of 2012. The Iron Dome rocket defense system had recently proved itself in combat along the Gaza Strip, and Israel was making impressive strides in other technological fields. So were Israel's enemies. That February, Iran independently launched its third satellite into space, and Hezbollah was continuing to amass missiles of unprecedented quantity and quality as neighboring Syria rocked in an endless civil war.

As veteran Israeli military correspondents, we were reporting daily on the fast pace of events and conflicts in Israel, along its borders and throughout the wider region. But we felt that part of the story was missing. Israel was in the midst of one of its largest military buildups in history, acquiring new drones, stealth aircraft, submarines, accurate missiles and better-protected tanks. Israel's enemies, primarily Iran and Hezbollah, were doing the same. It was an impressive arms race with potentially deadly consequences.

Over the years, the Israeli military has been a key focus of our attention. We are both veterans of the Israel Defense Forces (IDF) and continue to serve in the reserves. We have not only observed Israel's conflicts but have also become part of the story. The Second Intifada, the withdrawal from Lebanon, the pullout from Gaza, the Second Lebanon War and the various IDF operations in Gaza since are all events we have covered from the front lines and, in some cases, from behind enemy lines as well.

Our work has taken us on Israeli submarines and missile ships, on Israeli helicopters and C-130s and inside armored personnel carriers with IDF infantry troops on predawn raids in the Gaza Strip and the West Bank.

Over the years, we have closely followed the IDF's evolution. We watched as the military adapted to new threats—whether it was Palestinian suicide bombers, Hezbollah rockets or Iran's nuclear program. In recent years, the military has become even more vigilant, as it finds itself in a region in the throes of an unprecedented and historic upheaval. Known in the beginning as the hopeful "Arab Spring," this regional earthquake has given birth to new enemies like ISIS, now deployed along Israel's northern and southern borders.

Israel relies heavily on the reputation of deterrence it has worked hard to create over the years. We believe that this deterrence rests on three key pillars—Israel's purported nuclear weapons capability, its strategic alliance with the United States and the conventional capabilities of the IDF.

This book tells, for the first time, the story—of how Israel developed and created superior technology and weaponry for its military. It is a story that travels throughout Israel's history, from its start as a newly founded state up to today, as that state continues to face threats and challenges from across the region.

We decided that the best way to tell this story would be to

break up into separate chapters the accounts of the technology and weapons Israel excels at developing. For the most part, we have kept the story to a chronological timeline, but it also jumps—from the 1960s to today and then back to the 60s or 70s. We did this purposely, to give you, the reader, a complete picture of each weapon—how it was born, who the innovators were and what made them and their technology special. While each story is unique, the whole story is greater than the sum of its parts.

THE WEAPON WIZARDS

INTRODUCTION

"**P**ass me the binoculars," the Israel Defense Forces (IDF) chief of staff, Lieutenant General Benny Gantz, said to the officer next to him. He lifted the lenses to his eyes. Images from miles away came into sharp focus as he squinted under the winter sun.

Gantz was standing high atop Mount Kabir, doing something he loved: assessing his domain and exploring every inch of the country he was tasked with protecting.

He turned north and captured a clear view of the snowcapped peak of Mount Hermon along Israel's border with Syria. A quarter-pivot around to the east and he could scope out Jordan. Just below, a mere nod down with his binoculars, he was able to peer into the city of Nablus, home to about 130,000 Palestinians.

Those views provided a quick reminder of how small Israel really is. There is no such thing as strategic depth, Gantz stood there thinking. The enemy simply sits right up alongside us.

"What's that, over there?" Gantz asked Colonel Nimrod Aloni, the regional brigade commander who, like the chief of staff, had begun his military career in the paratroopers. "That," Gantz said, pointing, "the big white building with all the windows?"

Pushing aside his firearm, Aloni adjusted his own pair of binoculars. "Oh," he said. "That's the shopping mall."

Nablus is not just any Palestinian city. During the Palestinian unrest of 2000, known as the Second Intifada, Nablus had become the home for Israel's most wanted. At the time, Gantz had been the commander of the IDF division responsible for the West Bank. Terrorists from Islamic Jihad and Hamas set up bomb labs and headquarters throughout the twisty stone maze of Nablus's Casbah, or old city. Founded by the Romans and then built up by the Mamluks and the Turks, the Casbah was infamous for its network of tunnels and hiding places, convenient for a terrorist on the run.

IDF troops were frequently sent to raid the city and hunt down the terrorists. But in recent years, Nablus had been thriving. Terrorism was at an all-time low, and the IDF had significantly scaled back its incursions into the city.

Following the end of the Intifada, subsequent Israeli governments had tried to negotiate a peace deal with the Palestinians. Ehud Olmert had made a historic offer to Palestinian president Mahmoud Abbas in 2008, only to have it rejected. In 2009, Israeli prime minister Benjamin Netanyahu had agreed to freeze settlement construction in an effort to restart peace talks. That was an unprecedented move, but while talks ultimately restarted, they had once again failed to produce a deal.

At the time of Gantz's visit in 2012, the Palestinian Stock Exchange, based in Nablus, was hitting record highs as markets across the Arab world were stuck in the red. A new round of peace talks was expected soon, and hope was in the air.

But Gantz's West Bank tour had another purpose.

A couple of years earlier, what had begun as street protests in

Tunisia had spread like wildfire and given birth to what became known as the Arab Spring. Muammar Gaddafi was captured and executed in Libya, Hosni Mubarak was dramatically overthrown in Egypt and Bashar al-Assad was continuing to fight rebels in Syria, in a deadly, bloody and controversial war that would see the rise of ISIS and Global Jihad. In Lebanon, Hezbollah was continuing to amass sophisticated and advanced weapons, threatening Israel no longer as a guerilla organization but as a full-fledged military.

In Israeli defense circles, fears were mounting that the instability would spread, and Gantz wanted to make sure quiet would prevail in the West Bank and that, if it didn't, the IDF would be prepared.

Gantz had not originally been chosen to become the chief of staff but had attained the post quite arbitrarily, after the first candidate was deemed ineligible.

IDF chief of staff Lieutenant General Benny Gantz talks to soldiers during a military drill in 2011. IDF

So Gantz, summoned back to duty from retirement, had donned his uniform and assumed the lofty position.

"What I love most," he would tell people, "is being out in the field, with my soldiers."

After a few intelligence briefings, the tour came to an end. Gantz slid into the backseat of his armored jeep for the drive to a nearby helipad. His aide was already jittery. As usual on days like this, Gantz was running way behind schedule. The jeep left the base and turned onto a bumpy side road that wound around a nearby Jewish settlement, its stucco white homes and red roofs perched on the hill above.

"Stop the car," Gantz suddenly told the driver.

"What?" the driver asked, looking at the barren road in front of him, smack in the middle of the West Bank.

"I said, stop the car," the chief of staff repeated a bit more firmly. "Just pull over here."

The driver slammed the brakes and stopped next to an overpass.

"Get me Nimrod on the phone," the chief of staff told his aide, referring to the regional brigade commander who had accompanied him on the tour. Gantz took the phone.

"Nimrod, my jeep has been hit by a roadside bomb. I am injured and one of the soldiers with me has been abducted," he said, immediately hanging up. Aloni didn't even have a chance to respond.

Gantz climbed out of the jeep, looked at his silver Breitling watch and sat on a nearby rock. He picked up a twig, brushed the dust off and twirled it in his hands. "Now, we wait," he said.

Within minutes, the road was swarming with heavily armed soldiers, alerted to search for the "abducted" soldier. Armored Hummer jeeps, equipped with plasma computer screens showing the location of all nearby forces, took up positions on the hills above. A light buzz could be heard, coming from reconnaissance drones hovering in the skies above.

As the minutes ticked by, Gantz kept one eye on the soldiers and another on his watch. When Aloni finally showed up 10 minutes later, Gantz didn't have much to say.

"Okay, thanks. See you again soon," he said as he climbed back into his jeep, leaving behind a cloud of dust.

It was a regular workday for Gantz, but it presented an opportunity he seized to make a point. The Middle East was in the midst of great turmoil, and the IDF commander wanted to ensure his troops were prepared for a war that could erupt at a moment's notice.

"With this level of uncertainty in the region, we likely won't have the luxury of receiving a warning the next time war comes knocking," Gantz said. "We will win though, because our soldiers will be prepared and will have the best technology to assist them."

※ ※ ※

The equipment involved in what transpired that afternoon, during the surprise drill with Gantz, was just a microcosm of the military technology coming out of Israel and flooding the global arms market.

The plasma computer screens inside the Hummer jeeps that scrambled to the scene were part of Tzayad, the revolutionary command-and-control system used by the IDF. Tzayad, Hebrew for "hunter," works something like a GPS navigation system in a car but in this case displays the exact location of all forces in the area, while differentiating between friendly and enemy forces. If a soldier detects an enemy position, all he has to do is tap the location on the digital map and it will appear, immediately, on the screens of all other Tzayad users.

This technology is changing the ways wars are fought, with

obvious battlefield ramifications—shortening the time it takes to detect an enemy and lay down fire, otherwise known as the sensor-to-shooter cycle. Tzayad's accuracy and successful use in the IDF have been noted. In 2010, Australia paid $300 million for the system, and in 2014, a Latin American country bought it for $100 million.

The soldiers who scrambled to the scene to secure the IDF chief of staff during the abduction drill were carrying the Tavor, a new assault rifle developed by Israel Weapons Industries (IWI). Due to its light weight, high accuracy and short size, the Tavor has replaced the American M-16 as the IDF's weapon of choice. Since it entered service in Israel, the Tavor has reached every corner of the globe, from Colombia to Azerbaijan and Macedonia to Brazil.

Gantz's jeep, which was attacked in the imaginary bombing, was protected with armor designed and produced by Plasan Sasa, an Israeli company located in a small kibbutz in the Upper Galilee, along Israel's volatile border with Lebanon.

The company was founded in the 1980s among the kibbutz's white stucco homes and kiwi groves. It quickly gained the attention of the IDF with its innovative armor, made of dense composite material that could protect vehicles from rocket-propelled grenades (RPGs) and improvised explosive devices (IEDs) without adding significant weight.

When the US went to war in Afghanistan and then Iraq, IEDs soon became the greatest cause of fatalities. Orders at Plasan skyrocketed, and so did the company's profits, jumping from $23 million in 2003 to over $500 million in 2011.

In the skies above Nablus, IDF drones were keeping a close eye on the Palestinian city and the Israeli troops stationed nearby. Earlier that year, the IDF had launched the Sky Rider Program, under which it equipped field battalions with the lightweight

A Skylark drone is thrown by an IDF soldier during a military exercise in southern Israel in 2013.   IDF

Skylark drone, made by Elbit Systems, a leading Israeli defense contractor. Launched like a football thrown by a quarterback, the Skylark provides key over-the-hill intelligence, critical for infantry operations. Its delivery to the IDF continued to solidify Israel's standing as a world leader in the development of drones and unmanned systems.

From satellites to missile defense systems and drones to cyber warfare, Israel is at the forefront of new military technology being deployed on the modern battlefield. This book will tell the story of how Israel—a tiny nation of just eight million—has turned into one of the world's most prominent military superpowers and is developing technology that is changing the way wars are being fought around the globe.

Israel's success has led aerospace giants, weapons manufacturers and even countries to flock to the Jewish State to learn about this unique combination of innovation, drive and technology.

Large corporations in the US, France, the UK, India, Russia and Australia are regularly signing joint ventures with defense companies in Israel that are sometimes a fraction of the size of their foreign counterparts.

This story becomes even more compelling when one considers that just 60 years ago, Israel's main exports were oranges and false teeth. Today they are electronics, software and advanced medical devices.

According to *Jane's*, the British military trade publication, Israel is one of the world's top six arms exporters. Weaponry alone constitutes about 10 percent of the country's overall exports, and since 2007, Israel has exported about $6.5 billion annually in arms. In 2012, its 1,000 defense companies set a new record, exporting $7.5 billion worth of weaponry.[1]

Despite its small size, Israel invests more than any other country in Research and Development (R&D)—about 4.5 percent of the country's gross domestic product (GDP)—and continually tops lists as the world's most innovative country. While Israel's investment in R&D is impressive on its own, about 30 percent goes to products of a military nature. By comparison, only 2 percent of German R&D and 17 percent of US R&D is for the military.[2]

As popular newspaper columnist and CNN host Fareed Zakaria wrote of Israel: "Its weapons are far more sophisticated, often a generation ahead of those used by its adversaries. Israel's technology advantage has profound implications on the modern battlefield."[3]

✳ ✳ ✳

How did Israel do it?

This is the primary question this book will attempt to answer through stories of how Israel developed its unique weapons and tactics. Each weapon came of age in a different era and under different circumstances. The weapons' inventors were driven by different inspirations and motivations and drew on the country's different national characteristics, which together have created Israel's unique culture of innovation. No characteristic stands alone. They succeed all together in contributing to Israel's development as a military superpower.

Israel is often described as a culture of contradictions. It has tried making peace with the Palestinians for decades but has, until now, failed to repeat the success it had with Egypt in 1979 and Jordan in 1994. It has compulsory military service for men and women, but instead of instituting social discipline, the military is believed to be the primary source of the country's infamous casualness and informality.

Israel is a country of only eight million people and without natural resources, but is the country with the third-largest number of companies, after the US and China, listed on the NASDAQ. It has been engaged in a military conflict every decade since its establishment but nonetheless draws approximately three million tourists a year.

Part of the explanation for Israel's economic and military success has to do with the threat matrix the country faces and its nonstop battle for survival since its very inception.

General Gantz's mother, for example, was an inmate in the Bergen Belsen Nazi concentration camp during the Holocaust and came to Israel after the war. She was one of tens of thousands of Jewish refugees from postwar Europe searching for a new home. They were joined by hundreds of thousands of Sephardic

Jews who were forced to leave their homes in surrounding Arab countries after statehood was declared.

The fight for survival was nonstop. Throughout the country, food was rationed, public transportation was nonexistent and medical services were unreliable. Israel's existence was constantly in question.

When the War of Independence began in 1948, many of these Holocaust survivors were met at the docks, handed rifles and sent to fight on the front lines. They didn't know a word of Hebrew, and many were killed on the battlefield. But soldiers who fought alongside them told stories of their bravery and eagerness to finally be able to defend their people and homeland.

And while conditions were tough, the adversity Israelis faced from the outset forced them to develop critical tools—like the ability to improvise and adapt to changing realities—which they needed to survive.

"In this newly born IDF, the initial deficiencies in numbers, weaponry and training were compensated for by dedication and motivation, intelligence and improvisation," Reuven Gal, a former deputy Israeli national security advisor, explained. "These eventually came to personify the Israeli soldier."[4]

With barely any resources beyond the human capital that had immigrated to the new state, Israelis had to make the most of the little they had. Adversity and constant calls, beginning at the time of Israel's founding, for its destruction—even today, from places like Iran—foster creativity. In other words, if Israel is not creative in its thinking, there is a chance it will not survive.

This equation is a simple one, and as Haim Eshed, the man who came up with the idea for an Israeli satellite program, told us: "The shadow of the guillotine sharpens the mind."

But that can be only part of the answer. Israel is not the only country in the world that thrives in the face of adversity. South Korea, as an example, faces similar national security threats and has a fast-growing economy, but it lags behind in the development of advanced weaponry.

What makes Israel unique is the complete lack of structure. While this seems strange to cite as an advantage, it is exactly this breakdown in social hierarchy that helps spur innovation.

The absence of hierarchy is evident everywhere in Israel—in the military, on the streets and even in government offices, where low-level staffers call ministers by their nicknames.

And Israel is a country of nicknames—Prime Minister Benjamin Netanyahu is publicly called "Bibi." Former defense minister Moshe Ya'alon is referred to as "Bogie." Israel's president, Reuven Rivlin, is called "Ruvi," and the head of the Opposition, Isaac Herzog, goes by "Buji."

In everyday life, people in Israel look to cut social corners. Living in a small country with often just a single degree of separation from their leaders and other prominent individuals, Israelis excel in the use of *protexia*—the Polish word for "connections"—whether trying to get accepted into university or get an appointment with a well-known cardiologist.

As already mentioned, the mandatory service in the IDF is believed to be the primary source of this informality. In it, Israelis are imbued with a strong bias against hierarchy and a keen sense of chutzpah, the famous Yiddish word loosely translated as "audaciousness," "nerve" and "gall."

As new recruits, IDF soldiers are ordered to call their commanders "Sir." But after a few months, the commanders initiate a process called "Breaking the Distance," after which soldiers can call their commanders by their first names and are no longer obligated to salute them.

Think about this for a moment: the Israeli military, an entity expected to entrench structure and discipline in its soldiers, holds a ceremony to celebrate the demolition of hierarchies.

"This combination of informality and no hierarchy is the greatest advantage Israel has over other Western countries," Martin Van Creveld, the renowned military historian, told us. "Israel is relatively small and everyone knows everyone and since almost everyone serves in the IDF it's very easy to bridge the distance."

A culture of informality that lacks hierarchy may appear on the surface to endanger a country's or organization's ability to engage in long-term strategic thinking. In Israel, though, it has the opposite effect. Breaking down barriers creates an atmosphere that encourages and enables the free exchange of ideas. When officers of various ranks can engage at the same level and speak freely with one another, new ideas are fostered.

Take, for example, what happens when the commander of the Israeli Air Force (IAF) flies on a training mission. You would expect him to fly with senior pilots like himself. Instead, he usually takes the backseat to a younger pilot, sometimes half his age.

"There are no ranks inside the cockpit," Major General Ido Nehushtan, a former air force commander, told us after one such flight in an F-16 with a 25-year-old lieutenant.

The young learn from the old and vice versa. After these flights, the junior pilots can even criticize their superiors' performance without risk of demotion, loss of a promotion or any other punishment. They are actually encouraged to do so.

"This is the type of culture we work really hard to create," Nehushtan explained. "One of openness, professionalism and fairness."

For foreign officers who visit Israel, this culture often comes as a major shock.

That's exactly what happened when US Air Force lieutenant

general Ron Kadish, a former director of the United States Missile Defense Agency, made his first trip to Israel, in 1992.

Kadish served at the time as the US Air Force's F-16 program director. The number of F-16 aircraft crashing was on the rise, and Kadish had come to consult with the IAF, which possessed one of the largest F-16 fleets outside the US.

When he arrived at the base, his Israeli host took him on a tour of the different squadrons and showed off the aircraft, many of which had kill markings—small red circles with blue dots— from fighting in the First Lebanon War a decade earlier. One Israeli F-16 alone had shot down seven Syrian aircraft.

After the tour, Kadish was brought into the base commander's office for a technical discussion about the aircraft. There were the usual refreshments on the table—crispy and warm cheese and potato bourekas alongside thick and bitter Turkish coffee. Both sides presented their mechanical and technical assessments of the plane.

Then one of the participants started arguing with the base commander about the plane's drawbacks. Kadish asked the participant to identify himself. He was a noncommissioned officer, a lowly mechanic, who was arguing with a one-star general. Yet, he presented his case and was listened to, because ranks aside, he made sense.

"I sat there impressed," Kadish recalled. "In the US it is much more structured and people there need to be encouraged to state their minds. I didn't see that in the Israeli military in general and certainly not in the air force."

Kadish had just experienced a classic case of Israeli chutzpah. In the US military, speaking out of turn is unheard of, especially when it means arguing with your commander in the presence of a visiting foreign officer. In Israel, though, no one thinks in those terms. What the mechanic was doing was exactly what he had been trained to do and what he thought was expected of him—to speak his mind.

✳ ✳ ✳

In the Israeli reserves this attitude is emphasized even more. Officers who want to be promoted not only have to impress their superiors but also have to find favor in the eyes of their subordinates.

"If a reservist doesn't get the answer he wants he will go straight to the commander of the commander," retired brigadier general Shuki Ben-Anat, a former head of the IDF Reserves Corps, told us. "He doesn't do this to undermine the system but to get what he wants. Hierarchy doesn't mean anything to the reservist."

Colonel Shlomi Cohen, commander of the Alexondroni Brigade—one of the IDF's elite infantry reserves units—experienced this firsthand when he convened his soldiers for a debriefing after the Second Lebanon War in 2006.

When the war broke out, the Alexondroni reservists had enlisted in high numbers. Two soldiers had been abducted by Hezbollah, and rockets were falling throughout the home front. This was a war for survival.

But when the ceasefire kicked in and the reservists crossed back to Israel, their frustration and anger was too much to keep a lid on. They had been sent into Lebanon with outdated and defective equipment. They had to privately raise money to buy flak jackets and flashlights. They were also upset at the way Cohen had led them in battle. Orders were constantly changed and lacked decisiveness. Some days, they just sat inside southern Lebanese villages as if they were waiting for Hezbollah to attack. Resupplies never reached the reservists, forcing them to break into Lebanese grocery stores to scour for food. Some felt guilty and placed wads of cash on the counters as they left.

Two days after the war ended, Cohen convened his troops in an attempt to iron out some of the issues. They met in the pine tree forest just below the ancient city of Safed in the North, which had been struck repeatedly by Hezbollah rockets during the war. Cohen warned the reservists about the consequences of their complaining. At one point, he accused them of low motivation.

This was too much for the reservists. Some started yelling. Others began booing until Cohen finally got up and left. The reservists were infuriated, and several decided to take their protest to the Prime Minister's Office in Jerusalem.

The negative sentiment felt by Cohen's soldiers made its way up the chain of command, and instead of being promoted, the once-promising officer was sent to wrap up his career as Israel's military attaché in an Eastern European country.

Western militaries would recoil from the idea of booing a senior officer, but in Israel this was deemed acceptable. For the reservists, there was nothing strange about what they were doing. An officer had made mistakes, and they were upset. That he was their superior and they were still in uniform was a minor detail.

Ben-Anat recognized the reservists' feeling of frustration and disappointment. He had that same feeling in 1973 after the Yom Kippur War, which a state commission of inquiry found to be fraught with systematic failures and mistakes. While he was still in his compulsory service at the time, the war and its failures—at one point, Ben-Anat's company of just a handful of tanks was outnumbered 50 to 1—made him realize the importance of investing in the reserves corps and understand that Israel could not afford to be taken by surprise again.

After the war, he decided to continue serving, even though he had been officially discharged. While most reservists were called up for 14 to 21 days of service a year, Ben-Anat served 120 days, enabling him to climb the ranks even from the outside, and despite the fact that he worked for one of Israel's intelligence agencies. In 2008, after 35 years of service, he was bestowed with the rank of brigadier general and appointed the IDF's chief reserves officer.

Unlike some of its Western counterparts, the IDF relies heavily on reservists during times of war as well as for routine operations. This dependence dates back to the founding of the IDF as a "people's army" with a mandatory enlistment. While the objective for establishing a reserves corps was to ensure that there would be enough soldiers in emergencies, Ben-Anat claims that the presence of reservists has had a positive effect on military bureaucracy.

"Reservists come for a set period of time, and the last thing you want to do is make them feel like they are wasting their time," he explained. "This has an overall impact by making the entire system more effective."

Having a military based on a reserves force means that even after soldiers are discharged, go to university and enter the workforce, they continue to serve in the military every year. Pilots usually continue flying one day a week, while combat soldiers are drafted for two- to three-week stints each year—half for training and the other half for routine patrols and border operations.

This means that engineers who work for defense companies meet soldiers not just in boardroom meetings to look over new weapons designs, but also during reserves stints, when they themselves put on uniforms and become soldiers again.

Israeli engineers' experiences from the battlefield, as well as their continued training and combat in the reserves, help them better understand what the IDF requires for the next war as well as how to develop it. This means that any "operational require-ment" the military issues for a new weapon system is concise, clear and defined to the smallest detail. These people were in war, saw battle and know exactly what they need.

"We know what it means to sit in a military vehicle," an em-ployee from Plasan Sasa, the company that manufactures armor for IDF and US tanks, explained, "what it's like to hit an explo-sive device or take a burst of gunfire."[5] Those experiences are en-graved on one's mind.

"This almost firsthand familiarity between Israel's defense needs and what science and technology can deliver is unparal-leled in other countries," according to Dan Peled, a business pro-fessor at the University of Haifa.[6]

The US, for example, installs military officers in development teams at defense contractors, but they are often viewed as outsid-ers. In Israel, the outsiders are the insiders. Military experiences become lifelong experiences. This dual identity is a national asset.

Van Creveld put it more bluntly: "If 95 percent of your people never served in the military and were never in a military op-eration, how can you be expected to come up with innovative weapons?"

※ ※ ※

What also caught Lieutenant General Kadish's eye during his 1992 visit to Israel was the youth of the pilots and soldiers he met at the air force base, who were doing the work of American and European officers sometimes twice their age.

In the US Armed Forces, the average age is 29. In the IDF, it's slightly over 20.

What this means in practical terms is that junior Israeli officers and regular conscripts receive an immense amount of authority and responsibility at young ages. They also have fewer senior officers on top of them—the ratio of senior officers to combat troops in Israel is 1 to 9, while in the US it is 1 to 5—leaving the young soldiers with no choice but to make key decisions on their own.

In Israel, young intelligence analysts, after just a couple of years in the military, often have direct access to the defense minister and the prime minister. Twenty-three-year-old officers are made company commanders and given responsibility over sections of borders or parts of the West Bank. If terrorists infiltrate via their territories and carry out large-scale attacks, they are held responsible.

Giving responsibility to soldiers at such a young age contributes to their development as leaders, not just in the military but also later on in life. Since Israel is almost always in a state of conflict, its soldiers experience danger early on and are forced to make life-or-death decisions, sometimes on more than one occasion.

"Harvard graduates might get a first-class education and a doctorate but it is all theoretical," David Ivry, a former commander of the Israeli Air Force and director general of the Defense Ministry, told us. "In the IDF, soldiers get a doctorate in life."

Investing so much in soldiers has another result: they are viewed as priceless and as the children of all Israelis. Society acts accordingly. In 2011, Israel released more than 1,000 prisoners in exchange for a single soldier who had been held captive by Hamas in the Gaza Strip. Prisoner exchanges like these have been carried out by Israeli governments since the 1980s.

This is incomparable, not just in the Middle East but even among other Western militaries. The value placed on a single soldier makes every soldier important not just to his family, but to the entire nation. When a soldier is abducted, every family feels the pain, knowing that it could have been their loved one.

Mandatory service does something else to Israeli society: it serves as a melting pot. That's not the case in Western militaries, which are based on a system of volunteers. In the US, about a decade ago, 44 percent of soldiers were from rural areas, 41 percent were from the South, and nearly two-thirds were from counties where the average household income was below the national median.[7]

In Israel, almost everyone serves. Men are drafted for three years and women for two. A rich kid from Tel Aviv who is drafted into a combat unit will find himself or herself training alongside an Ethiopian Jew from a development town in the South, a Russian immigrant from the North and a religious soldier from a

New recruits to the IDF's Golani Brigade train in March 2016.   IDF

settlement in the West Bank. Service in the IDF doesn't tolerate social barriers. Poor Israeli kids who never would have had an opportunity to operate sophisticated technology get that chance in the IDF. Kids who grew up without a smartphone at home are suddenly trained to become cyber operators. When citizens are in uniform, socioeconomic and racial labels are ripped away.

This melting pot is part of the recipe for fostering innovation. Creativity can happen only when people come together and exchange ideas. To do that, they need to know each other and share the same language and culture. In Israel, they do that in the army.

The IDF also encourages its officers to get a "multidisciplinary education." This stems from the limited resources Israel has at its disposal, in terms of not just raw materials but also people. In foreign aerospace companies, Israelis like to joke, there are experts for a single bolt or fuse. In Israel, engineers cross over to other fields and specialize in more than one task.

That is why many senior officers and top executives in Israeli defense companies have different degrees in different fields. An IDF officer, for example, is encouraged to get a BA in electronics and then an MA in something else, like physics or public policy.

Brigadier General Danny Gold, the mastermind behind the development of the revolutionary Iron Dome rocket defense system, is a good example. He took a sabbatical in the middle of his air force career and received two doctorates—one in business management and the other in electrical engineering. As we will show, he needed both to get the revolutionary Iron Dome off the ground.

If there is one unit that best represents the IDF's investment in manpower and the focus it puts on multidisciplinary education, it is Talpiot, the place where Israel's best and brightest serve.

Talpiot—the word comes from a verse in the Song of Songs and refers to a castle fortification—is Israel's premier technological unit. Every year, thousands try out but only about 30 get accepted, a privilege that entails signing on for nine years of service, three times the usual length.

These soldiers usually have skill sets that make them suitable to be pilots or operators in elite commando units. But Talpiot trumps everyone and takes whoever it wants. It is that important.

The unit was born out of a disaster, the 1973 Yom Kippur War. Israel was caught unprepared when Syria and Egypt attacked on the holy Jewish fast day. More than 2,000 soldiers were killed, and countless aircraft and tanks were destroyed. If until then Israel thought it had a superior military, it now had a sense of vulnerability not felt since the country was founded a quarter of a century earlier.

While Israel ultimately held on to its territory, the traumatic war was a stark reminder that innovative tactics were not enough to retain military superiority. Israel needed a technological edge. The question was how to get it.

Shortly after the war, Colonel Aharon Beth-Halachmi, head of the air force's Technology Department at the time, received a phone call from Shaul Yatziv, a physicist at Hebrew University he had met earlier that year when visiting the university to see a high-powered laser Yatziv was developing. The Soviets and Americans were working on lasers as well, and Beth-Halachmi thought the IDF should invest in a similar capability. Military applications could be figured out later.

Yatziv said he had something important to discuss and that he would be bringing along a friend. A few days later, he showed up at Beth-Halachmi's office with Felix Dothan, another physicist. Beth-Halachmi felt like he was meeting a modern-day version of the biblical Moses and Aaron—like Moses, Dothan had difficulty speaking. Yatziv served as his spokesman.

Dothan, Yatziv said, had written a paper proposing the establishment of an institute he called "Talpiot." The program, they said, would be strictly for Israel's geniuses. These soldiers would go through a 40-month training program—the longest in the IDF—and each would receive a degree in physics, mathematics or computer science while completing combat training with the elite paratroopers.

The graduates would then spend time in each of the military's different branches. At the end of the 40 months, they would be posted to a single unit, with an emphasis on the air force or the Intelligence Corps.

Beth-Halachmi was intrigued. He, too, was distressed by Israel's performance in the war and was looking for ways to improve the IDF's technological capabilities. He agreed to take the proposal to his superiors.

What made Talpiot unique was its focus. Instead of being taught one skill, participants would receive a multidisciplinary education and become familiar with the entire spectrum of the IDF's technological capabilities. The idea was to provide them with skills needed to come up with solutions that cross bureaucratic borders and technological limits.

But not everyone was thrilled by the idea. Officers in the air force and military intelligence opposed the program. They wanted the best recruits to serve as pilots and field commanders. "It would be a waste to send them somewhere else," was the typical reaction Beth-Halachmi heard throughout the General Staff. There was little he could do from his position in the air force. He would have to wait.

A couple of years later, Beth-Halachmi was promoted head of the IDF's Research and Development Authority and got his own seat around the General Staff table. This meant that he had open access to Chief of Staff Raful Eitan. At one of their weekly meetings, Beth-Halachmi presented the Talpiot idea, and Raful was sold. He didn't even bother convening a meeting. Within three months, a pilot phase was launched.

It didn't take long for Beth-Halachmi to notice that the program was a success. A few years after its establishment, the prime minister convened a special meeting of the Israeli Security Cabinet to discuss the program. A few generals were complaining that the graduates were not being distributed fairly throughout the military's different branches. Everyone, including Israel's spy agencies, wanted a "Talpion," as the graduates are called. It was a tough meeting, following which the prime minister ruled that Talpions needed to be assigned to all of the country's different security agencies, including the police. Nowadays, there is an average of five units competing for a single Talpion.

"What we showed was that you don't need a lot of people for breakthroughs," Beth-Halachmi told us. "All we needed were the right people with the right training."

The success stories are innumerable and, for the most part, remain classified. One Talpion invented a way for projectiles to travel 10 times their regular speed, by propelling them with electric and not chemical energy.

Another Talpion, who turned down medical school to enlist in the unit, invented a new seat for helicopter pilots. During his military service in the late 1980s, the Talpion learned that a high number of pilots were suffering from back pain. So he designed a new seat, installed it in a helicopter simulator, cut a hole in its backrest and trained a pen on the pilot's back. Then, using a high-speed camera, he recorded the effect the vibrations were having on the pilot's back.[8]

Another Talpion played a key role in developing a system to

detect cross-border terror tunnels being dug into Israel along its border with the Gaza Strip.

While Talpiot may be a small unit—it has produced only about 1,000 graduates in some four decades—its impact is felt throughout the entire IDF and beyond. Graduates have found their way into the upper echelons of Israeli academia and the country's high-tech industry, founding and taking up top posts in dozens of companies, many listed on the NASDAQ.

"There is no other program like this in the world," said Evyatar Matanya, a former Talpion who later became head of Israel's National Cyber Bureau. "A Talpion often revolutionizes a unit singlehandedly. Two or three in one unit is already a different world."[9]

✳ ✳ ✳

We believe the secret to Israel's success is a combination of all of the above but also runs deeper, into the core of Israel's national character.

Few other countries in the world, if any, have been embroiled in conflict for as long or as intensively as Israel. There is little margin for error when the enemy you are fighting is just a few minutes' drive from your front door—when the terror groups along your borders regularly fire rockets at your homes and schools and send suicide bombers onto your buses.

In this reality, security is never taken for granted. Some Israelis get nervous after a long period of quiet. It can't be real, they say. It must be the quiet before the storm.

Israel was the first Western country to fight against Soviet military machines in Egypt and Syria and the first modern state to face suicide terror on its streets, years before New York or London, Madrid and other capitals in Europe. From a possible military strike against Iran, to the occasional manhunt for a terror suspect in the West Bank, Israel faces more threats than most countries and is constantly developing state-of the-art military technology to deal with them.

"What works to our benefit is the combination of three elements," Udi Shani, a former Defense Ministry director general, told us when we met in Tel Aviv. "We have: innovative people, combat experience to know what we need and immediate operational use for what we develop since we are almost always in a state of conflict."

But while Israel's development of weaponry is revolutionizing modern warfare, it has taken place not in a vacuum but, rather, in the Middle East, possibly the world's most volatile region. Israel may view its need for cutting-edge weaponry as a reaction to external threats, but it is exactly this technological prowess that often fuels the exact arms race it is trying to prevent.

In 2010, for example, Israel launched Stuxnet—one of the first known military cyber attacks in the world. The Israeli computer virus was so effective that it destroyed around 1,000 centrifuges at Iran's main uranium enrichment facility and set back the country's illicit nuclear program by nearly two years, according to some estimates. Since then, though, Iran has set up its own cyber unit, investing over $1 billion annually in creating effective offensive capabilities. A full-fledged cyber war appears to now be only a matter of time.

As instability spreads throughout the Middle East, and more countries, particularly in Europe, face urban terror threats from ISIS and other terror groups, the tactics and technology perfected by Israel are in high demand.

The Iron Dome short-range rocket defense system, for example, has helped Israel turn a strategic threat—rocket fire from the

Gaza Strip—into a manageable tactical problem. This allows Israeli leaders to stay focused on the larger challenges and threats their country faces.

The Trophy active protection system, installed on IDF Merkava tanks and capable of intercepting incoming RPGs and antitank missiles, enables these big steel fighting machines to remain relevant in an era of asymmetric and urban warfare. At a time when most countries are phasing out their armored corps, Israel is doing the exact opposite.

Israel as a story has always marveled the world. It is a tale of how a weak and ancient people returned to their homeland, established a state and, against all odds, not only survived but prospered.

This book will add a new layer to that story. It is not just about the technology that has brought Israel victory and success on the battlefield; it will also zoom in on the people and unique Israeli culture that made this possible.

In a world full of uncertainty and danger, this story is one we should all pay attention to.

1

BEGINNING IN A BUNKER

It was 1945, three years before the State of Israel's establishment, and the Jewish leadership in Palestine already had a sense of what was looming. It was a matter of time before the Mandate was dissolved and the British left Palestine. The moment they did, the Jews knew, the Arabs would attack.

Weapons were in short supply, but the real problem was that under British rule, Jews caught with weapons faced prison and sometimes the death penalty. The Haganah—the Jewish paramilitary organization that would eventually evolve into the IDF—needed ammunition and weapons. The question was how to get them.

The man tasked with finding a solution was Yosef Avidar, a senior Haganah commander. Avidar had been born in Russia and undergone basic military training at the age of nine, with the help of a non-Jewish neighbor who had returned home from service in the Czar's army. The skills stuck for life, and after arriving in Israel, Avidar stood out and quickly climbed the Haganah ranks.

He was commander of Jewish forces in Jerusalem's Old City during the 1929 Arab riots and succeeded in holding off Arab attackers with a single gun and a mere 11 bullets. In a cable he later sent Haganah headquarters, he criticized what he deemed to have been a gratuitous use of ammo. His complaint points to just how little ammunition the Jewish community had.

"We could have stopped them with seven bullets, "Avidar said. "We wasted four."

The riots were a wake-up call for Avidar, who understood that if Israel was going to survive, the Jews needed training and lots of it. That's how he found himself one Saturday morning at an amphitheater near the Hebrew University campus on Mount Scopus, teaching a group of 50 men how to throw grenades. The Haganah didn't have real grenades, so Avidar held the demonstration with a homemade contraband one. As he lifted his hand to throw the grenade, it exploded prematurely, ripping into his flesh. The explosion could be heard for miles, and it was only a matter of time before British forces arrived. Despite the serious injury, Avidar refused to be evacuated until all his men had safely snuck away.

That was a defining moment. The Jewish community needed quality weapons, the kind that wouldn't explode in soldiers' hands. But the British had a tight grip on the country. Smuggling Jewish immigrants in by sea was hard enough, let alone weapons. As a result, Avidar came up with a revolutionary idea to build a bullet factory in Israel, the first of its kind. It was a bold notion. First, the Yishuv—as the Jewish community in pre-state Israel was known—didn't really have any experience in manufacturing weapons. Second, the British were everywhere. It would be difficult to set up a bullet factory without their finding out.

But Avidar was determined. Success, he knew, depended largely on finding the right location. He toured the country and settled on a hilltop just outside the city of Rehovot, known as the home to the Weizmann Institute of Science, one of the country's leading academic institutions. A small group of Jews had settled on the hilltop in 1932 but eventually moved into the city to consolidate forces in the face of growing Arab attacks.

The hilltop had two clear advantages—it was isolated but close enough to the city and power grid. Since it was elevated, there was room to dig into the mountain to build the factory underground, far from the eyes of the British army.

Avidar also liked the fact that the hilltop was close to the Rehovot train station, which was always full of British soldiers. The last place the British would expect a weapons factory, he figured, would be right under their noses.

But Avidar needed a cover story to explain why a group of Jews was suddenly settling on that exact hilltop. He was told that a group of new immigrants—affiliated with the Jewish Scouts movement—was planning to establish a kibbutz. One day, Avidar showed up in their dining hall and asked the group to alter their plans a bit. Instead of establishing a kibbutz, Avidar suggested they join the war effort. They would move to the hilltop, the Haganah would build all of the necessary structures and they would work inside the factory. The group agreed.

By spring, a few dozen 20-year-olds had moved to the hilltop and begun what seemed like ordinary lives among the citrus orchards and communal recreational activities.

At the same time, construction began on the underground weapons factory, named the Ayalon Institute. Some of the existing structures were renovated—the bathrooms, the chicken coop, the kitchen and the cafeteria.

To build the underground hall, Avidar recruited a Jerusalem-based contractor who had participated in the construction of Hebrew University on Mount Scopus in the 1920s, one of the most grandiose construction projects at the time and the same place where Avidar had severely injured his hands a few years back. Within 22 days, the contractor dug out a 100-foot-long hall some 30 feet underground.

To anyone who asked, the pioneers said they were building the underground hall to store fruits and vegetables they would be picking in the nearby orchards and fields. The hall, they explained, was needed to keep the produce fresh.

The secret hall was covered by a thick concrete ceiling with two openings, each leading into a newly built structure: one to a bakery, and the other to a laundry. A production line was assembled below with World War I–era equipment, purchased in Warsaw and smuggled into Israel via Beirut. The copper needed for the bullet casings was smuggled into Palestine in crates marked as containing lipstick cases.

To cover the noise the bullet machines would be making, Avidar needed the laundry to work around the clock, and to do so, it needed customers. So Avidar had the group open a branch in downtown Rehovot, which soon began handling washing for most of the region. The laundry business won a bid for a nearby hospital and later took on as a customer the British army, the exact entity it had been established to dupe.

No chances could be taken. Sunlamps were installed in the underground hall to ensure that the "kibbutzniks" making the bullets looked tanned, as if they had been out in the fields all day.

The opening in the bakery was covered by a massive 10-ton stone oven, built on rails so it could slide on and off a secret stairwell. The opening in the laundry was covered by the washing machine, which also moved with the pull of a lever.

Avidar, who would go on to become one of the IDF's first generals and an Israeli ambassador to the Soviet Union, secretly connected the factory directly to the nearby power grid to prevent anyone from asking questions about why a small, new kibbutz was using so much electricity.

And finally, to discourage visitors, kibbutz members spread rumors in town that their community had been struck by an outbreak of hoof-and-mouth disease. They put up a sign at the kibbutz gate, ordering visitors to dip their shoes in disinfectant before entering.

Women work in the bullet factory built underneath a laundry known as the Ayalon Institute.   AYALON INSTITUTE

The plan worked brilliantly, and the British never suspected a thing. There were a couple of close calls, though. One was in early 1948, when a train carrying British soldiers from the Gaza Strip to Lod was derailed by a mine explosion just below the kibbutz. Twenty-eight soldiers were killed and dozens more were injured. The attack was carried out by the Lehi, a Zionist group also known as the Stern Gang, which was more radical and militant than the Haganah.

The immediate concern was that the British, suspecting that the kibbutz had played a role in the bomb attack, would search the buildings. Work was suspended in the subterranean bullet-production hall, and all of the workers were ordered outside.

But how would they prevent the British from searching the kibbutz? The workers decided to rush to the train and offer help—food, water and medical care. The British naturally assumed that there was no way the same kibbutz that so graciously offered assistance would also be the kibbutz that had carried out the attack.

The Ayalon Institute operated for almost three years—from 1945 until the establishment of the State of Israel, in 1948—and manufactured more than two million nine-millimeter bullets. At their peak, factory workers manufactured 40,000 bullets a day embossed with two letters: E for "Eretz Yisrael" (Hebrew for "the Land of Israel") and A for "Ayalon."

✳ ✳ ✳

After the war, the Ayalon Institute folded into Israel Military Industries (IMI), the country's first defense company and today a recognized global leader in the development of missiles, rockets and armor.

But that would take some time. In the meantime, the state-to-be needed a way to get weapons, and with war coming, it needed them fast. The problem was that almost no one was willing to sell arms to the beleaguered soon-to-be state—not the United States, Britain or the Soviet Union.

The one exception was Czechoslovakia.

The first planes that served the Israeli Air Force were four Messerschmitt aircrafts gathered from Nazi Luftwaffe leftovers in Czechoslovakia. Each plane was taken apart, shipped to Israel, reassembled and equipped with a machine gun and four 70-kilogram bombs.

Other planes, brought in through Italy, had fuel tanks installed in place of seats to extend their range and enable them to make the flight to Israel.

The Czechs also agreed to supply Israel with rifles and four artillery guns, last used in World War I. It didn't make a difference. If it could shoot, the Israelis wanted it.

Beyond the arms deals, there were also shady and creative plots hatched to obtain weaponry. One group of Israeli arms buyers went to England and set up a fake movie company to film what they claimed would be a World War II movie. They hired an entire cast and crew—actors, producers—and even purchased aircraft to be used in the movie.

In one of the opening scenes of an air battle, the aircraft took off into the cloudy London sky. The cameras filmed from below as the planes flew away and then as they turned southeast, toward Israel.

These gimmicks were nice, but Israel's leaders knew that they couldn't go on forever. Israel needed to find a more secure way to get weapons. That would have to wait, though. First, Israel needed to fight for its survival.

In May 1948, as expected, war broke out. In a coordinated assault, five Arab armies from across the Middle East invaded Israel. The new state didn't seem to stand a chance. The Arabs had superior arms—tanks, artillery cannons and an organized air force. The Israelis didn't have a single cannon or tank.

Estimates regarding Israel's chances varied, but defeat was a definite possibility. In a briefing to the Jewish leadership, one top military commander gave the new state a 50-50 chance of survival.

"We are as likely to win as we are to be defeated," the commander said.[1]

The war was brutal. Israel was outnumbered and outgunned. More than 6,000 Israelis were killed, and another 15,000 were wounded. In the end, though, the Jewish State survived, thanks to a combination of unconventional tactics, amazing dedication and never-seen-before improvisation. Israel achieved the impossible.

One example was in Yad Mordecai, a small kibbutz on the Mediterranean coast just north of the Gaza Strip, where some 150 Israelis fended off an entire Egyptian mechanized division for six days with just 75 rifles, 300 grenades and a single anti-tank rocket launcher.

Then there was the story of Lou Lenart.

Born in Hungary in 1921, Lenart immigrated with his family to the United States at the age of nine and became a regular target for anti-Semitic beatings and taunts in the small Pennsylvania town where they settled. He learned early in life that to survive, he needed to be strong, and to be strong, he needed to join the Marines.

Lenart became a pilot, and by the end of his seven years of service, he had flown combat missions in the Battle of Okinawa as well as against the Japanese mainland. After the war, he learned that his relatives, who had remained in Hungary, had been killed in Auschwitz. Lenart returned to Los Angeles and began thinking about Israel, or, as it was then known, Palestine.

"My family had been killed in Auschwitz and I felt that the remnants of the Holocaust had a right to life and some happiness—and no one wanted them except their own people in Israel," he recalled for us years later.

He joined up with a group of Jewish pilots in Los Angeles that was being formed to help Israel. He arrived in the country in April 1948, just a month before war would break out. He was immediately put to work reassembling some Czech Avia S-199 Mule combat aircraft.

By mid-May, war had erupted, and the planes were finally ready. After about a week of fighting, Israel was on the verge of despair. A column of 15,000 Egyptian soldiers with 500 vehicles and tanks was stopped on the Mediterranean coastal road near Ashdod, just a few miles south of Tel Aviv. Israeli soldiers had blown a bridge the night before, but the Egyptians were hours away from repairing it. If they did, they would be in Tel Aviv by the morning. If Tel Aviv fell, Israel would be lost.

Lenart heard about the stalled Egyptian column and gathered his pilots. They were going to fly south to bomb the Egyptians, he told them. The problem was twofold: The planes had just been assembled and had not yet participated in real missions, and it was not 100 percent guaranteed that they would work. The other problem was that the existence of the planes was still a secret. This was not the way the country had planned on introducing its new air force to the world.

But that wasn't going to stop the pilots. The stakes were too high. As the formation leader, Lenart flew in first over the Egyptian column. Diving down over a group of vehicles, he let his bombs loose and luckily hit a fuel truck, setting off a series of secondary explosions. The other pilots followed suit and together began strafing the ground troops with machine-gun fire.

The Egyptians were taken by surprise, and after several hours, they gave up trying to reach Tel Aviv and turned east to join the Jordanian battle near Jerusalem. Tel Aviv was saved. The bridge where the Egyptians were stopped would later be named Ad Halom, Hebrew for "until here."

Lenart didn't have time to consider the consequences of that successful operation. Anyhow, he told people, it didn't make a difference if it was luck, destiny or superior instinct. There was still a war that needed to be won.

❋ ❋ ❋

"We have a unique military problem," the "Old Man" said. "We are few and our enemies are many."

It was 1953. One war was over, but the "Old Man"—David Ben-Gurion, Israel's first prime minister and the Jewish State's version of George Washington—knew that it was just a matter of time before the next one.

Ben-Gurion was troubled by the basic question of how Israel, a tiny country of just a few hundred thousand Jews surrounded by millions of hostile Arabs, would continue to survive.

So he took a vacation from work in Jerusalem and traveled to the modest cabin he kept in Sde Boker, a small kibbutz perched over the Ramon Crater, a natural wonder in the southern Negev Desert.

A few days later he returned to the Prime Minister's Office in Jerusalem with a paper entitled "The Doctrine of Defense and State of Armed Forces," which to this day, with some minor changes, continues to serve as the basis for Israel's national defense.[2]

The rationale was simple and remains true for Israel in the modern age: Israel needed a strong qualitative military edge.

Since Israel had fewer soldiers than Syria, it needed to have better-trained ones; since it had fewer tanks than Egypt, it needed more advanced ones; and because it would ultimately purchase the same F-15s as Saudi Arabia, the Israeli Air Force's version of the aircraft needed to come equipped with specially designed smart bombs as well as advanced electronic warfare systems. Israel simply needed to make sure that it always had superior-quality weapons and fighters. Not necessarily more of them; just better ones.

The question, though, was how to accomplish that goal. No one imagined then that this new, resource-poor country could establish and maintain independent research, development and production capabilities.[3]

Therefore, Ben-Gurion concluded, Israel needed to find countries willing to sell it arms. In 1950, though, that option was a nonstarter. The United States, Britain and France had issued the Tripartite Declaration, a multinational pledge to suspend arms sales to the Middle East. If arms were supplied to Israel, the countries thought, the Soviets would supply weapons to the Arab states. If they didn't, then the Soviets would also hold back.

This left Israel with one real option: to go rogue and search for weapons through an assortment of schemes and adventures, sometimes joining hands with the most unlikely of partners.

To lead the effort, Ben-Gurion placed the fate of the country in the hands of a 26-year-old Polish-born kibbutznik by the name of Shimon Perski, who had Hebraized his last name to "Peres."

Peres had impressed Ben-Gurion when serving as his aide during the War of Independence, and he was the kind of person who the Israeli leader believed could attend fancy diplomatic cocktail dinners but later, at night, roll up his sleeves and load illegally purchased weapons into shipping containers at the docks. Peres was appointed deputy director of the Defense Ministry delegation to New York, where his mission was clear—find and procure arms for the IDF.

Peres's escapades took him to meetings with shady characters like Jimmy Hoffa, the notorious Teamsters leader, and to places like Bogota, Colombia, where he made an emergency landing in a plane whose engines were in flames. He negotiated a deal to purchase 46 tanks from Mexico and even received the keys. Only later did Peres discover that the tanks didn't exist, having disappeared somewhere along the Mexican border.

As he traveled, Peres learned that no one really cared much about Israel's problems. Like many great salespeople, he discovered that the secret to getting what he wanted was to make his partners understand why it was in their interest to cooperate—in this case, to work with the new and isolated Jewish State.

In Havana, Peres tried to persuade top officials in the Cuban secret service to purchase weapons on behalf of the beleaguered nascent Jewish State. Peres showed up for the meeting at police headquarters at 12:00 noon—the time he thought he had been summoned—only to be told by a giggling secretary that his host, Señor Efraimo, never held meetings during the day.

"[The] Señor meant midnight," the receptionist explained.

Peres came back to the police headquarters 12 hours later, but instead of being invited into an office, he was taken by his Cuban host to a local nightclub. After some drinks and a bit of flirting with the local girls, Peres and Efraimo got down to business.

"What could we do," Peres later explained. "We had to conclude many of our deals with dark figures, even with gangsters. We had no other way."[4]

A year later, Peres learned that Canada was selling surplus

World War II cannons, of the same kind the IDF desperately needed. A quick inquiry revealed that Canada was willing to sell Israel the cannons but wanted $2 million, money Israel didn't have.

Peres decided to try to raise the money. He traveled to Montreal and met with Samuel Bronfman, a well-known Jewish philanthropist and owner of one of the largest liquor companies in the world.

Bronfman agreed to speak with the Canadian government and try to negotiate a discount. Once he did—reducing the price by $1 million—he asked Peres who was going to give him the other million.

"You," the Israeli replied.

Bronfman was a bit taken aback by the Israeli's brash response but quickly regained his composure and called his wife to dictate a list of 50 guests who needed to be invited for dinner that night. Then he looked over at Peres and noticed that he was wearing a blue suit with white socks. "You can't come to dinner with such socks," Bronfman said, making his guest stop at a department store on the way home to buy dress socks.

Later that night, as the main dish was being served, Peres gave his pitch about the cannons and how vital they were for the IDF and Israel's survival. The guests opened their checkbooks.

This gung-ho culture was unique for Israel, a country that had been founded against all odds and was continuing to survive against all odds. What Peres was doing—cutting corners and smuggling weapons—could be expected. With the future of the Jewish State on the line, almost anything was legitimate.

While in New York, Peres reconnected with an old friend by the name of Al Schwimmer, a Jewish flight engineer who helped smuggle aircraft into Israel before the War of Independence, including the Czech combat planes Lenart had used to bomb the Egyptians.

Peres had gotten to know Schwimmer during the war, when he served at Ben-Gurion's side, and was taken even then by the American's vision, deep conviction and loyalty to his people.

Schwimmer had worked for TWA, but when World War II broke out, he enlisted in the US Air Force and spent the war flying across the Atlantic Ocean more than 200 times. Being Jewish did not mean much for Schwimmer at the time, but a visit to a liberated concentration camp and a meeting with a group of Holocaust survivors sparked a burning desire to help the Jews achieve independence in their historic homeland. Jews, he believed, could be safe only if they had their own country.

When he returned to the States, Schwimmer tracked down the Haganah's representative in New York and offered his services. It took a while, but the Haganah came back with a clear but risky request—help us build an air force. This wasn't exactly legal. According to the Neutrality Act, US citizens were forbidden from exporting, without government approval, arms to countries at war.

But Schwimmer was determined to help. As a military veteran, he was entitled to purchase surplus aircraft at discounted prices. He gathered a group of Jewish pilots and engineers with whom he had served during World War II, and together they began buying up aircraft—anything they could get their hands on that could fly. Very few of them knew the real reason, and most were told that Schwimmer was helping Panama establish a national airline to transport cattle to Europe.

The planes were taken to a hangar near Los Angeles, where they were repaired, disassembled, loaded into crates and shipped or flown to Italy. There, the Haganah had located an abandoned airfield and was gathering aircraft before the flight to Israel.

After Israel's War of Independence, Schwimmer returned to the US despite criminal charges waiting for him. He and his crew rented a mansion owned by Jeanette MacDonald, a Hollywood star known for a series of 1930 musicals.

Schwimmer stood trial and was convicted but got off without jail time. He was fined $10,000, stripped of his voting rights and veteran benefits and barred from holding a federal job. While he never sought a pardon, he received one in 2001 from outgoing president Bill Clinton.

The conviction didn't slow Schwimmer down. He quickly got back into the smuggling business, largely with Peres's encouragement. As a front, he set up a new business called International Airways. The office was located in a corner of Lockheed Martin's factory in Burbank, north of Los Angeles.

One of Peres and Schwimmer's first joint operations was the smuggling of Mustang combat aircraft to Israel. The US Air Force had retired the single-seat bombers but refused to sell them to Israel. Instead, the planes were sold to a Texas junkyard. What the authorities didn't know was that the owner immediately flipped them to Schwimmer for the same price he had paid.

When the planes arrived in Burbank, Schwimmer and his crew reconstructed them to make sure they were operational and then disassembled them again, packed them in crates marked "refrigeration equipment" and shipped them to Israel.

By 1951, the group was secretly sending Mosquito light bombers from the US to Israel. Some of the planes were disassembled and shipped to Israel in containers. Others were flown directly to Tel Aviv, albeit with a few refueling stops along the way. On one of the smuggling missions, a plane went missing over the Canadian province of Newfoundland.

The pilot was Ray Kurtz, an American Jew from Brooklyn who had served in the US Air Force during World War II. After the war, Kurtz became a fireman at Engine Company 250 on Foster Avenue in Brooklyn but quit his job in 1947 to join Schwimmer's illicit plane-smuggling scheme.

In one memorable operation, the longest-range bombing mission carried out by Israel at the time, Kurtz flew a B-17 bomber from a Czech air force base all the way to Cairo. The bombs were slated for Abdeen Palace, one of the Egyptian president's official residences, but they missed and scattered nearby. Even so, the bombing raid was a huge feather in the new state's cap: Israel had succeeded in penetrating deep into Egypt.

Kurtz, a hero in Israel, went missing while smuggling more aircraft to the state. Peres and Schwimmer decided to immediately launch a search mission and set up a base of operations in the subarctic town of Goose Bay.

The search was based on information collected from local Eskimos who claimed to have seen the Mosquito dive into the snow. While the rescue teams spent seven days flying over the glaciers and mountains, they found nothing.

Despite the failed mission, an idea that would change Israel was born in Goose Bay. During the long polar nights, Peres and Schwimmer spent hours talking and dreaming about the day Israel would no longer need to rely on clandestine schemes to get its aircraft. One day, Peres declared prophetically, Israel would have its own aerospace company and build its own airplanes.

Peres remembered the looks of pity he received from most of the rescue team. They thought he was hallucinating. How could a tiny country that had to fly contraband aircraft over the Arctic expect to one day build its own planes? But Schwimmer took him seriously and assured Peres that it was possible. "At Goose Bay," Peres later said, "the Israeli aircraft industry was founded."[5]

When they returned to New York, Peres heard that Ben-Gurion, who was in the US for an official state visit, had just arrived in California. Together with Schwimmer, he boarded a plane to report to the Old Man on the failed rescue mission.

Schwimmer and Peres met Ben-Gurion at the Lockheed Martin facility in Burbank and brought him to Schwimmer's corner repair shop. Ben-Gurion walked through with a puzzled look. He didn't understand how Schwimmer was able to repair aircraft with so little equipment. "Why don't you come to Israel?" the prime minister asked. "We need our own aerospace industry. We need independence."

For Ben-Gurion, an independent aerospace industry was exactly what he meant when speaking years earlier of a qualitative military edge. An Israeli aerospace company would ensure aerial dominance over the entire Middle East.

It took some convincing, but Schwimmer understood what was at stake. He accepted Ben-Gurion's offer with a few conditions—the company would have to be free of cronyism and would work like an American corporation.

Ben Gurion agreed. "We need you in Israel," he said. "Come."

Within a week, Schwimmer had drafted a 30-page work plan listing every piece of equipment he would need to get started—from hydraulic cranes to an assortment of bolts and screws.

Now came the hard part—getting the government to agree to finance the project. Peres and Schwimmer flew to Israel and started promoting their idea—to establish Israel Aerospace Industries (IAI)—in a series of meetings with top government and military officials.

As expected, there was resistance from the start. The head of the air force announced that Israel didn't need an aerospace company; the Finance Ministry refused to provide a budget, and the transportation minister refused to even think about it. Israel hadn't produced its own cars, he said. How could it even think about building planes?

Peres refused to give up. He succeeded in raising some money, supplementing it with additional funds from the defense budget. Within a few months, construction on a new hangar began on the outskirts of the city of Lod, adjacent to the country's international airport. Schwimmer flew back to the US to purchase the necessary equipment.

By 1955, the company was up and running. Flying Fortresses, Dakotas, Mosquitos, Stearmans and other aircraft—purchased from anyone willing to sell—were brought in for repairs.

Within a year, the company had more than 1,000 employees. By the mid-1960s, it had 10,000 workers and was Israel's largest single employer.

Israel Aerospace Industries didn't stick to just repairing aircraft. By 1960 it was rolling out independently manufactured combat aircraft, built according to blueprints obtained in France. That huge milestone gave IAI the confidence to seek new and more complex challenges. An Israeli company was on its way to becoming an international superpower.

At the end of 1951, Peres returned to Israel. Ben Gurion was impressed with the results his protégé had shown in New York and appointed him director general of the Defense Ministry, the most senior nonpolitical civilian position in the Israeli defense establishment. While IAI had been established, it was just getting started. Israel still had no way of getting weapons on its own. It needed a regular supplier. It needed a country.

"We should cast as many fishing lines as we can," Peres told his aides at the time. "Perhaps something will bite."[6]

With the Tripartite Declaration still in effect, though, Israel's options were limited. But then, in 1955, everything changed. The Soviet Union decided to supply Egypt with $250 million worth of modern and sophisticated arms via Czechoslovakia, the same country that had supported Israel before the War of Independence. A potential conflict was brewing, and the West realized that it could no longer sit on the sidelines.

In Jerusalem, the news hit hard. The Egyptian-Soviet deal included advanced MiG fighter jets, long-range bombers and hundreds of tanks and armored personnel carriers. Israel needed someone to turn to. One idea, pushed by Moshe Sharett, the Israeli prime minister at the time, was to ask the US for help. America had been the first country to recognize Israel at the United Nations, and as the home of about five million Jews, it had close cultural ties with the Jewish State.

Sharett believed that if Israel followed a policy of greater restraint, it might be able to persuade the US to supply weapons equal in quality and quantity to those the Soviets were supplying Egypt.

The problem was that the US wasn't budging from the policy, adopted by President Eisenhower, not to become a major arms supplier in the Middle East. The Americans would give Israel economic aid but not weapons.

Then there was England, which Pinchas Lavon, the defense minister at the time, believed was a viable candidate to become Israel's main arms supplier. Peres went to London but was met with a cold shoulder. Upset over a recent IDF raid in Gaza, the British refused to deliver some tanks Israel had ordered earlier that year. England had also already agreed to sell two naval destroyers to Egypt and didn't appear ready to switch sides.

This left one viable option—France. Aside from England, France was then the only European country that manufactured most of its weaponry domestically, including fighter jets, tanks and artillery cannons.

At the time, Israel didn't really have a defense relationship with Paris. It purchased a bit of weaponry through a color-

ful Polish count who lived in a Parisian palace and served as
Israel's front, but there was no official connection. So Peres flew
to Paris and set up a meeting with the deputy prime minister.
Within a few weeks, he had successfully negotiated the purchase
of 155-millimeter cannons.

Peres arrived in Paris during one of the republic's most chaotic
periods. Governments were rising and falling one after another,
and Israel feared that the instability would make it impossible to
establish a clear and strategic partnership with France. What some
saw as a weakness, Peres viewed as an opportunity. He realized
that with instability and disorder, he could navigate among the different government offices and establish the necessary personal ties
needed to make deals happen, no matter which party was in charge.

*Shimon Peres (on the right) and David Ben-Gurion visit the nuclear reactor as
it was being constructed in southern Israel in the early 1960s.*   MOD

Peres was also banking on the sympathy he sensed for Israel within French defense circles. In the 1950s, France was fighting in Algeria to maintain control over the North African country. The Arabs, including Egypt, supported the Algerian rebels. This scenario was a classic case of what Peres thought of as "my enemy's enemy is my friend."

While Peres was pounding the pavement in Paris, in Israel there was a feeling that doom was on the horizon. By the beginning of 1956, the Soviet arms had started to reach Egypt. Jerusalem announced an "Israel Deal," urging citizens to volunteer their time to strengthening fortifications throughout the country. People showed up at the Defense Ministry with jewelry and watches, anything that could boost the arms fund. There was a sense that the moment Egyptian president Gamal Abdel Nasser had his weapons, he would attack.

It would take numerous trips, but Peres eventually succeeded in establishing the necessary ties in France, and despite an occasional setback, the arms began to flow.

Israel's ties with France quickly evolved beyond arms sales. In July 1956, Nasser announced the nationalization of the Suez Canal. Peres recognized an opportunity and decided to use the crisis to push the Israeli-Franco alliance to the next level—convincing Paris to sell Israel a nuclear reactor.

Within a day, Peres had a meeting with French defense minister Maurice Bourgès-Maunoury, who wanted to know how long it would take Israel to cross the Sinai and retake control of the canal. When Bourgès-Maunoury asked if Israel would be willing to join the French and British in a tripartite military operation, Peres seized on the opportunity. "Under certain circumstances, I assume that we would be so prepared," Peres responded.

The "circumstances" would become Israel's most prized possession—the sale of a nuclear reactor that would be built in the desert town of Dimona and would provide Israel with an un-matched level of deterrence in the Middle East. The French agreed, and a few weeks later, leaders from France, the UK and Israel met in Sèvres to finalize the invasion plans. Right before signing the agreement, Peres met privately with the French prime minister and defense minister.

"It was here that I finalized with these two leaders an agree-ment for the building of a nuclear reactor at Dimona, in southern Israel, and the supply of natural uranium to fuel it," Peres later said.[7]

Later that month, the IDF invaded Egypt. The Mustangs Schwimmer and Peres had smuggled to Israel a few years earlier flew in first, with a special cable connected to their fuselage to cut Egyptian communication lines. That brilliant move caused mas-sive confusion among Egyptian forces in the Sinai. After a couple of days, the IDF was joined by the French and British. The war did not go as planned but Israel had achieved its goal of further solidifying its relationship with France.

By 1958, arms were flowing regularly between Israel and France, and the countries were discussing the possible sale of advanced fighter jets. It seemed like the balance of power was being restored to the Middle East.

Israel wanted to deepen the ties. So the Defense Ministry asked if one of its air force pilots could enroll in the French test pilot course. Paris agreed and rolled out the red carpet.

The pilot selected for the course was Danny Shapira, an up-and-coming aviator born and bred in the Land of Israel. Shapira had fallen in love with aviation as a child when he saw the *Graf Zeppelin* German passenger airship fly over his hometown of Haifa.

By the age of 15, Shapira was sneaking off to the cinema to see any movie made with fighter pilots, planes or even a tidbit about aviation. He was sailing hang gliders and sinking his teeth into anything aeronautical at the boys' club in a nearby kibbutz. In 1944, at the age of 19, Shapira got his pilot's license.

In May 1948, as the Jewish community was preparing for war, Shapira was sent with a group of pilots to Czechoslovakia to complete combat training. The group left Israel on May 13, on the last civilian airliner allowed to depart Palestine before Ben-Gurion's declaration of independence the very next day. This was the IAF's first pilots' course.

By 1959 Shapira had made a name for himself as one of the country's best fighter pilots, flying bravely during the War of Independence and the Suez Crisis.

The air force was desperately in need of its own test pilot, so the IAF commander, Major General Ezer Weizman, sent Shapira to France to take the advanced test pilot course. Shapira passed with flying colors. Afterward, Weizman asked that he remain in France to evaluate a new combat aircraft, called the Mirage.

Under development by Dassault, a French aircraft manufacturer, the plane was still top secret, but the French had built two prototypes and were interested in scoring a big contract with the IAF. The Mirage was a technological breakthrough at the time. It had a triangular wing and was the first European-designed aircraft capable of reaching Mach 2 speeds, thanks to its unique rocket propulsion system.

Weizman told the French that while he was interested in the plane, he would need one of his own pilots to inspect and fly it. The French argued, claiming that only French pilots were qualified to fly on the Mirage.

With typical Israeli chutzpah, Weizman told Dassault CEO Benno-Claude Vallières, "I sent Danny Shapira to your test pilot school and you gave him a certificate that says he can fly all of your different airplanes so let him! If Danny Shapira doesn't fly, there's no deal."

The French agreed.

The first flight went smoothly, and Shapira was impressed by the plane's capabilities. But then came the real test, in front of Weizman, on June 26, 1959.

The day of the flight, Shapira arrived at the base early to suit up. Mirage pilots had to wear special pressurized suits and oxygen masks. The pressure was immense, and only as Shapira crossed the runway to board the plane did he realize that he had forgotten to change into combat boots. It didn't make a difference. It was time for takeoff. The first part of the flight went smoothly, and Shapira easily climbed to 40,000 feet. He then pushed the rocket switch. The sudden increase in speed caught Shapira unprepared.

"I'm at Mach 1.1, Mach 1.3, Mach 1.9, Mach 2," Shapira reported back to the control room, where Weizman was watching with uncontrollable excitement. "Mach 2 for the first time in 2,000 years," Weizman shouted to the Dassault executives, a reference to the 2,000 years it took the Jewish people to reestablish a sovereign state in the Land of Israel.

But while Weizman was jumping for joy, Shapira was preparing for the worst. The plane kept climbing. At one point, Shapira looked over his shoulder and said to himself, "Danny, you're leaving earth."

Suddenly a red light went off in the cockpit, a sign that Shapira was exceeding the plane's maximum speeds. At 64,000 feet, he got the rocket engine to shut down, but he slowed too fast, and the plane began to shake. Shapira thought the fuselage would come apart. Ultimately, he succeeded in regaining control of the plane and adapted to the fragility of Mach speed, landing safely back at the Dassault airfield.

Though momentarily shaken, Shapira was sold on the plane, and so was Weizman. His next challenge was getting Dassault to modify the specs on the plane to suit Israel's needs. The Mirage was built as a high-altitude interceptor and was not designed to carry bombs for ground attacks.

Israel, though, needed aircraft that could be versatile—wage dogfights against enemy aircraft and bomb ground targets, all on the same mission. The French had built the plane to intercept So-viet bombers but conceded to Shapira that it could be reconfig-ured to carry air-to-ground bombs as well.

"We'll also need a cannon installed on the fuselage," he told the French engineers.

They were baffled by the request. "Passé," Shapira was told. "You don't need the cannon. That's for old planes."

But Shapira, who would go on to become an Israeli icon on a par with the likes of, in America, Chuck Yeager, refused to give in. He knew that when flying in skirmishes against Egyptian and Syrian aircraft, the distances were not enough to use an air-to-air missile. Israeli Air Force pilots needed a weapon that could fire at close range. They needed a cannon.

Shapira insisted, and the French eventually caved in, fitting the aircraft with twin 30-millimeter cannons. Danny Shapira had demonstrated typical Israeli chutzpah. He, a lone Israeli Air Force pilot, had persuaded one of the largest and most successful de-fense conglomerates in the world to redesign its fighter jet, just because he said so.

The first consignment of new Mirages arrived in Israel in 1962 and saw action just a few years later, during the Six Day War, when they blew 51 aircraft out of the sky, all by cannon fire.

✳ ✳ ✳

Israel's love affair with France would come to an end after the 1967 war. Charles De Gaulle, the French war hero and president, used the Six Day War as an excuse to cut ties with Israel as part of an effort to restore relations with the Arab world. He imposed a strict arms embargo on the sale of any weaponry to the Middle East and particularly to Israel.

Under President Lyndon Johnson, the United States eventually replaced France as Israel's main supplier of modern weaponry.

But as deep and powerful as this new relationship with the US would become, the French experience taught Israel an important lesson: to survive, the Jewish State could not rely solely on foreign assistance. It needed to find a way to develop its own R&D and production capabilities. It was a matter of survival.

2

CREATIVE DRONES

"The agent is back," the analyst said as he popped his head into Shabtai Brill's office. "He has pictures."

The year was 1968, and in the hyper-intensive intelligence atmosphere of the time, this was big news. Brill, a major in IDF Military Intelligence Directorate—known by its Hebrew acronym, Aman—set aside the report he was reading and got up.

Brill was used to seeing classified intelligence, but today was special. The "agent" was one of the first Israeli spies to successfully infiltrate Egypt since the Six Day War had ended, a year earlier. He had photos that were supposed to reveal Egyptian war plans, including possible preparations behind the cease-fire line.

A small crowd surrounded the agent in the department's main nerve center, a room where all intelligence flowed before being distributed to the various case officers. Colonel Avraham Arnan, Brill's direct superior, was focusing on one photograph.

"What do you think it is?" he asked the group of analysts. "It looks like a military bridge."

It was, and Egypt had moved the bridge to less than a mile from the Suez Canal, the strategic waterway that connected the world of commerce but separated Egypt from the territory it had lost to Israel during the Six Day War. The bridge could be used by tanks and armored personnel carriers to cross the canal and invade Israel. It was way too close for comfort.

Before sending the agent to Egypt, Israel had pursued other avenues to gather intelligence on what Egypt was doing just over the canal. One officer designed a special platform to mount on tanks so that intelligence officers could stand on them and peer over the 30-foot-high sand barriers the Egyptians had erected on their side of the Suez. The platforms seemed effective until the day an Egyptian sniper took a shot at one of them.

Next, the Israel Air Force flew reconnaissance aircraft along the border and took pictures of what was happening on the ground. But because of Egyptian surface-to-air missiles, the aircraft had to fly at high altitudes, rendering the pictures of little or no value. That left the IDF with one viable alternative—live agents on the ground in Egypt, passing for Egyptians and traveling to the Suez Canal via Europe to take photos of what was happening along the border.

Arnan took the photo down the hall to alert Aman's top brass. Brill stood there thinking how crazy it was that one single photo held the key to Israel's survival.

"We need to launch such an operation to get a single photo of what is happening just over the canal?" Brill asked. He could grasp the significance of the intelligence, but something felt wrong. It just didn't make sense that there wasn't an easier way to see what was happening a few hundred feet away.

On his drive home that evening, Brill couldn't shake the feeling that there had to be an easier way to gather intelligence over the canal. He recalled a movie he had seen a few weeks earlier in

Tel Aviv. The feature film was preceded by a short newsreel that included a scene about an American Jewish boy who had received a toy airplane as a gift for his bar mitzvah. The plot was not important once Brill's imagination started going. He remembered that the planes came in different colors, were wireless and pilotless, and could be flown by remote control. What Brill conceived seemed almost too easy: buy a few remote control airplanes, attach cameras to their bellies and fly them over the Suez to photograph Egyptian military positions.

Brill knew he would need partners to implement his idea. So he went to air force headquarters, snooped around and discovered Shlomo Barak, an officer who spent his weekends flying remote control airplanes. He was one of a handful of people in Israel at the time who had the necessary experience for what Brill had in mind.

Brill tried to get the air force to assume responsibility for the idea. They were naysayers. "Remote control planes are toys, and we have no use for them," officers from the air force's technology branch told Brill.

So he went back to his own commander and tried to sell the idea. "We can buy a few of these planes for real cheap, install cameras and fly them over the Suez to spy on the Egyptians," Brill told Arnan. But Arnan wasn't convinced. He first asked to see the planes in action.

Later that week, they met at a small airstrip outside Tel Aviv for a flight demonstration. Barak piloted the remote control plane, did some maneuvers, a flip or two, and landed it flawlessly. Arnan liked the idea but wanted to know what it would cost. Brill didn't know offhand, so, together with Barak, he compiled a list: three airplanes, six remote controls, five engines, a few spare tires and propellers. The grand total was $850.

Arnan approved the budget, and a member of Israel's defense

delegation in New York went to a Manhattan toy store, purchased the equipment and sent it back to Israel in the embassy's diplomatic pouch. This way, no one would question why an Israeli was traveling with so many toy airplanes in his luggage.

After their safe arrival in Israel, the planes were brought to the Intelligence Directorate's technological team for further development. They were fitted with 35-millimeter German-made cameras with timers programmed to take pictures automatically every 10 seconds.

"We're ready to go operational," Brill told Arnan a few weeks after the planes arrived. The senior officer was still skeptical. He feared the planes would be shot down by Egyptian anti-aircraft fire and suggested that they first see if IDF anti-aircraft teams could shoot them down.

On a hot summer day, Arnan and Brill drove down to the IDF's anti-aircraft training base in the Negev Desert, restricted one of the roads so it could serve as a runway and even gave the anti-aircraft gunners a heads-up as to the direction from which the planes would be flying. Brill was nervous. If a plane got shot down, it would mean the end of his idea.

The plane took off and started circling over a patch of sand, and the gunners opened fire. The sound was deafening, lasting what seemed like a lifetime. Brill lost sight of the plane and feared the worst. To his surprise, after the smoke cleared, the toy airplane was still there, soaring above. Barak tested flights at 1,000 feet, 700 feet and then even at a mere 300 feet. The gunners, though, could not make a successful hit. The toy airplane was too small a target. After the plane landed, the astonished Arnan turned to Brill and gave him permission to take the plane for flights over Egypt.

The first target chosen was a row of Egyptian military posi-

An unidentified soldier assembles the toy airplane Israel flew over the Suez Canal in 1969. SHABTAI BRILL

tions located near Ismalia, a town along the Suez and next to Lake Tismah, otherwise known as Crocodile Lake. The team chosen to fly the plane consisted of two people, one a "pilot" who operated the remote control and the other a "navigator," who watched it through a set of 120 x 20 binoculars and ensured that the pilot did not lose his line of sight.

The dramatic first flight, in July 1969, didn't go as smoothly as planned. First, since there were potholes everywhere, it was difficult to find a piece of road that could function as a runway. After the discovery of a 100-foot flat strip, takeoff was finally approved. Arnan gave permission to penetrate about a mile into Egypt. But then, when the plane was airborne, it entered a cloud of sand. Its momentary disappearance triggered panic that the plane would crash in Egypt and Israel's new secret weapon would be discovered. Barak, who served as the navigator, told the pilot to fly the plane in circles and to increase altitude. "Don't be pressured. Just keep flying until we see it," Barak told him.

After a few tense moments, the plane finally emerged from the cloud, and the pilot managed to land back in Israel. The film was immediately taken to be developed, and when the photographs came back, Arnan and Brill were stunned. The resolution was amazing. They could clearly see the trenches the Egyptian military had built along the canal. Even communication cables connecting the different positions were visible.

For the first time, Israel had clear photos of the obstacles the Egyptians were building along the Suez and how they were preparing for a future war.

After another mission, this one over the Sinai, Arnan sent the team to the Jordan Valley, where similar flights were conducted over Jordanian positions. The success was mesmerizing, and by the end of the summer, Major General Aharon Yariv, head of military intelligence, had decided to establish an official development team to build a small but sturdier remote control airplane that could be integrated into regular service. Yariv sent Brill a letter thanking him for his invention: "You deserve praise for this invention because without innovation at all levels and ranks, there would be no IDF."

A few weeks later, Brill was promoted and put in command of all early-warning intelligence systems in the Sinai. He was confident that he had left his pet project in good hands. It was time to move on. One day, some months later, he received a phone call from one of his original partners. Aman, he was told, was terminating the project. The team appointed by Yariv had tried to build a new airplane, instead of relying on existing platforms, and it

A photo of an Egyptian port along the Suez Canal, taken during the first flight of Shabtai Brill's toy airplane.   SHABTAI BRILL

kept crashing. As a result, Aman's top brass decided the project was too expensive and, anyhow, should be overseen by the air force. Aman was shutting the project down.

Brill refused to go down without a fight. Through the course of 1969, he sent a number of letters to Yariv and the rest of the country's intelligence brass and warned of devastating consequences should the project be abandoned. He pleaded with his commanders not to end the project. They refused to listen.

On October 6, 1973, on Yom Kippur, the Egyptian military launched a surprise and successful attack across the Suez, proceeding practically unopposed up through the Sinai Peninsula. While Israel ultimately held on to the territory during the bloody debacle, when the war ended the country was left in a state of trauma. More than 2,000 soldiers had been killed, the most since Israel's War of Independence.

Brill could barely contain his anger. He was certain that if his project had not been canceled, Israel would have detected Egyptian military movements and had time to bolster defenses or even prevent the war. Seeing what was happening just over the border could have saved thousands of lives.

"Had we continued taking pictures of what was happening just three miles over the canal we would have seen the Egyptian tanks, bridges and equipment amassing and understood they were preparing for war," he said. "Unfortunately, this didn't happen."

Aman understood its mistake, dusted off Brill's old plans and reached out to local defense companies to begin designing an Israeli lightweight unmanned aerial vehicle (UAV)—what today is more commonly referred to as a drone.

It would take another few years for the Israeli design to become operational, but in the meantime two things were clear: Israel needed quality intelligence, and that meant getting into the drone business. Brill could not have known at the time, but what he started on the shores of the Suez Canal in 1969 would burgeon one day into a massive, billion-dollar industry for Israel and position it as a global military superpower.

After several years of research, development and test flights, Israel's first drone—the Scout—was finally delivered to the air force in 1979. The first version of the Scout was launched by a rocket, but soon enough, state-owned Israel Aerospace Industries upgraded the model so it could take off and land on a runway, just like an airplane.

Almost immediately, the Scout was engaged in combat.

It was June 1982, and Israel had decided to invade Lebanon to end the rising cross-border terror and rocket attacks by the PLO.

The greatest obstacle was the presence of nearly 20 Syrian Soviet-made surface-to-air missile (SAM) batteries deployed in Lebanon's Bekaa Valley. The SAMs severely limited the air force's ability to maneuver.

The IAF had been preparing for war. In the weeks before, Scout drones flew over the valley to collect radar and communication frequencies from the SAM batteries. This was precious data needed for what the IAF planned to do next: electronically neutralize the batteries.

Israel's full-force attack was launched on June 6. An electronic warfare system succeeded in blinding and neutralizing most of the missile systems, and the Scouts assisted Israeli fighter jets in identifying and bombing the missile batteries. The operation was a major success. The IAF destroyed almost all of the Syrian SAMs and in one fell swoop knocked 82 Syrian MiGs out of the sky without losing a single Israeli fighter jet.

That operation caused a shift in Israeli thinking. Officers who until then had refused to believe in these new unmanned aircraft had a change of heart. The potential of these miniature drones suddenly seemed unlimited.

In the meantime, while Israel's Scouts were moving from one successful operation to the next, Israel's greatest ally, the United States, was having difficulty getting its own drones off the ground. Billions of dollars were being poured into projects that closed down one after another. Nothing seemed to work.

A few years earlier, the Pentagon had funded the development of Aquila, a drone built by Lockheed Martin that required a few dozen people for takeoff but kept crashing. In 1987, after burning through over $1 billion, the Pentagon decided to shut the program down.[1]

Boeing was also working on a drone—the Condor—that came with a 200-foot wingspan, double that of the U-2 manned reconnaissance aircraft it was being developed to replace. That program was also shut down after a $300 million investment. Only one Condor was built; today it hangs in a museum in California.[2]

In December 1983, the US finally decided to ask Israel for help. A few weeks earlier, the US Navy had launched a botched attack against Syrian anti-aircraft batteries stationed near Beirut in response to the downing of an American spy plane. The attack was a disaster: two American planes were shot down, a pilot was killed and a navigator was captured. While a few Syrian guns were destroyed, the Syrian anti-aircraft fire forced the US planes to drop their bombs far from their targets. An inquiry into the botched raid concluded that a nearby US battleship had cannons in range of the Syrian air defense systems and that they could have been used without endangering American pilots. The problem was that the navy had no way of knowing where the Syrian missile systems were located. It needed eyes in the sky to direct them.

A few weeks after the botched operation, Secretary of the Navy John Lehman traveled to Beirut and decided to use the occasion to fly to Tel Aviv to learn about Israel's use of drones. He had heard about the Scouts' use in 1982 but had never seen them up close. When he arrived at Israeli military headquarters, Lehman was taken into an operations room and asked to sit in front of a small TV. He was handed a joystick and given control over a drone in flight. Similarly, Marine Corps commandant General P. X. Kelley visited Israel to view the drone program. At the end of his trip, he was presented with a kind of home video, shot by a circling drone. In some of the footage, Kelley's head was fixed in the camera's crosshairs.[3]

Both men were sold. The next stage, though, was to figure out how to push the deal through the complicated US bureaucracy. Lehman decided to simply skip over the usual procedures and had the navy contract Israel Aerospace Industries directly to develop a new drone based on the Scout. The Americans wanted something bigger and stronger, with a more advanced avionics system that could serve as a spotter for battleships. IAI soon had a prototype, which it called the Pioneer. After a flight demonstration in the Mojave Desert, the US Navy was hooked. It ordered 175.

Delivery of the Pioneers started in 1986, and it didn't take long before they engaged in combat. In 1991, Saddam Hussein's Iraq invaded Kuwait. The US went to war to free the Gulf state. During one operation, a Pioneer drone flew over a group of Iraqi soldiers, who saw the aircraft and, not knowing what it was, took off their white undershirts and waved them in the air. It was the first time in history that a military unit surrendered to a robot.

A few months after returning to the US, Lehman learned of another drone under development in Los Angeles, which he was told could also potentially serve as a spotter for navy gunships. This drone was the work of an Israeli engineer who had recently left a senior management position at Israel Aerospace Industries— manufacturer of the US Pioneer—to try his luck in the US.

Born in Baghdad in 1937, Abe Karem had moved to Israel just after the state was established, in 1948. By the time he was eight, Karem knew he wanted to be an engineer, and a few years later, he found his true love—aviation. At 14, he built his first airplane, and within two years, was an instructor in his high school's toy plane club. After high school, Karem went to study aeronautics at the Technion, Israel's equivalent of MIT. He then joined the air force, and after his discharge he went to work for IAI.[4]

During the Yom Kippur War, in 1973, Karem built his first unmanned aircraft. The IAF was having difficulty penetrating Egypt's Soviet air defense systems, so within a couple of weeks, Karem's team had developed a decoy—basically a missile that could be flown with a joystick—that the IAF could use to activate the Egyptian radars, detect their location and then hit them with anti-radiation missiles fired from nearby fighter jets. Despite its success, after the war, the IAF decided to buy similar decoys from the US. Karem's version was buried. He argued for the importance of investing in domestic systems to create a local industry but failed. Frustrated, he quit and decided to try his luck in the US.

Karem and his family moved to Los Angeles but couldn't afford to buy a house and rent office space for Karem's new business at the same time. So Karem and his wife reached an uneasy but workable compromise: the house would be for the family, and the attached garage would be for the drones. Within months, Karem had set up Leading Systems in his 600-square-foot garage in Hacienda Heights, and together with two part-time employees, he began building a new drone. The idea was to build a drone as cheaply as possible, a lesson he had learned at IAI. Called Amber, the prototype was made of plywood and fiberglass and came with a two-stroke engine that Karem pulled out of a go-kart.

By the mid-1980s, Amber drones, still in their test phase, were flying daily, sometimes for as long as 30 hours straight. At Lehman's insistence, the navy announced plans to buy 200 aircraft. Karem thought he had made it. But then, in 1987, Congress cut the funding. Karem refused to give up and started designing a version of the Amber for export. But then the bank called; it was time to pay back a $5 million loan. Without money, Karem had no choice but to sell Leading Systems to Hughes Aircraft, which quickly flipped it to General Atomics. Karem stayed on as a consultant and developed a variant of the Amber called the Gnat 750.

The turning point for Karem came from a combination of the most unlikely of places—Bosnia and Israel. In 1993, ethnic war broke out in the former Yugoslavia; the combatants—soldiers and militiamen—wore civilian clothes, and the US government was encountering difficulty in assessing the situation on the ground.

The problem was brought to the attention of R. James Woolsey, the CIA director at the time. During a brainstorming session one day at Langley, Woolsey recalled a trip he had made to Israel a few years earlier, during which he had, for the first time, seen drones in action. While he was there, some Defense officials whom Woolsey knew from his previous post, as undersecretary of the navy, took him on a tour of a new drone unit the IDF had established. The unit was responsible for surveillance missions over Lebanon. An IDF colonel showed him around the base and introduced him to the drone operators.

"These folks really know what they are doing but they look awfully young," Woolsey told the colonel. The Israeli officer grinned. "This is the Israeli Model Airplane Club," he said. "We just drafted them all into the same unit."

The colonel then took Woolsey inside a nearby tent and showed him a video from a recent operation. On the screen, Woolsey watched a convoy of three Mercedes sedans drive on a road in southern Lebanon. Intelligence, his host explained, had identified a passenger in the second car as a senior Hezbollah operative. The drone, the officer continued, "lit up" the car with a laser target designator, enabling a nearby IAF helicopter to fire a missile and destroy it.

Woolsey had seen this use of laser guidance—referred to as "lasing"—when he served as general counsel for the Senate Armed Services Committee during the Vietnam War. Back then, fighter jets did the lasing, but Woolsey had positive recollections of the accurate airstrikes that followed. The tour came to an end and Woolsey went home, but he had been bitten by the drone bug.

Now at the CIA and facing a major intelligence gap in Bosnia,

Woolsey knew what was needed: "We need a long endurance drone," he told his staff. He summoned the Pentagon drone team to Langley and asked how long it would take to get a drone up over Bosnia. "We can do it," the Pentagon officials said, "but it will take six years and $500 million." That was too long and too expensive.

He then remembered Abe Karem. Woolsey had met the Israeli engineer a few years earlier and had been impressed by his innovative thinking. He called Karem and got straight to the point.

"How much would it cost and how long would it be before you could be up and operating over Bosnia?" he asked.

"Six months and $5 million," Karem said.

"Well, you are two orders of magnitude below the Pentagon," the CIA chief said. "Let's see what we can do."

Woolsey teamed Karem up with Jane, a CIA employee (whose full name cannot be published) who had developed a special command-and-control system for drones. Karem and Jane got to work, and after six months, the Gnat 750 was flying reconnaissance missions over Bosnia. A few days later, a live feed from the drones was installed in Woolsey's seventh-floor office at Langley, and the CIA director was able to watch foot traffic over a bridge in Mostar while communicating with the ground station through an early form of chat software.[5]

The Pentagon was impressed and immediately positioned itself to capitalize on the Gnat's success. It awarded General Atomics a contract to develop a more robust drone based on the Gnat with a bigger engine and new set of wings.

The biggest change to the Gnat was General Atomics' decision to place a satellite communication link on the aircraft. The company decided that the more advanced drone needed a new name, so it held a competition. The winner was "Predator." The drone would go on to become infamous as America's most lethal weapon in the global war on terror, responsible for countless strikes in Pakistan, Afghanistan, Iraq and Yemen. It took Israel and an Israeli engineer to help make that happen.

What makes drones appealing for militaries is that they can successfully carry out "3D" missions—dull, dirty and dangerous. "Dull" refers to routine, mundane missions like patrols along borders or maritime surveillance of seas and oceans. These are physically demanding and are extremely tedious and repetitive. While humans tire after 10 or 12 hours, the Heron drone—the Israeli Air Force's main workhorse since 2005—can stay airborne for 50 hours.

"Dirty" involves entering airspace infected by chemical or biological agents. While a human would have to wear cumbersome protective gear, drones can operate risk-free, making them more versatile. And "dangerous"? That's more open for interpretation but basically covers missions that can be done by a robot instead of a pilot who could be injured or killed.

Drones have an almost endless list of advantages, which make them preferable to manned combat aircraft. They are smaller, lighter, and cost less and can hover over targets for longer. Fighter jets have the advantage and disadvantage that they can break the sound barrier, and while speed is an advantage in a dogfight or a mission that requires a quick in-and-out, it means that the aircraft's presence can be identified almost immediately. Drones can hover over targets while their engines' humming noise blends into city traffic. This makes them the perfect weapons to hunt and eliminate moving targets, such as terrorists.

Since the delivery of the Scout, in 1979, the Israeli Air Force has used and retired a number of different drones. But unlike the larger fighter jets, attack helicopters and transport aircraft that are purchased overseas, Israel's drones are strictly blue and white, developed and manufactured by home-grown Israeli companies.

Since 1985, Israel has been the largest exporter of drones in the world, responsible for 60 percent of the global market, trailed by the US, whose market share is just 23.9 percent.[6] The customers have been dozens of different countries, including the United States, Russia, South Korea, Australia, France, Germany and Brazil. In 2010, for example, five NATO countries were flying Israeli drones in Afghanistan.

In today's IDF, drones are used by all military branches. The air force, for example, maintains drones like the Heron for reconnaissance missions on all of its various fronts—Gaza, Lebanon and Syria.

With a length of about 27 feet, the Heron is just a bit shorter than a Cessna light aircraft, although its wingspan is significantly longer, by about 20 feet. It is powered by a rear propeller, emitting a steady lawnmower-like sound. Its single best quality is its autonomous flight system, allowing the operators to insert a flight route before takeoff and then get the aircraft off the ground by pressing just four buttons. The drone then flies to its target and can be programmed to return to a predesignated point at the end of the mission. This allows the operator to focus on the mission instead of on flying the plane.

Heron's manufacturer, IAI, does not publicly divulge the drone's exact cost, but industry estimates put the price tag at approximately $10–$15 million, far less than the cost of a manned combat aircraft. For the price of one F-35 fifth-generation multirole fighter, one of the most recent IAF purchases, the air force can buy about 10 Herons.

The Heron can fly in two different modes, line-of-sight or satellite. The operator must be located within 250 miles of the drone at all times if it flies in line-of-sight mode. In satellite mode, the drone is controlled via a satellite linkup, meaning that distance is limited only by the amount of fuel it can carry. But the real significance of a drone is in its payload. Herons, for example, carry their cargo in more than one space—in their bellies, on their wings and in rotating gimbals mounted under the nose. The gimbals include the sensors, which vary based on the mission—day/night cameras, infrared vision, laser targeting as well as special sensors to identify weapons of mass destruction (WMD).

One Israeli-designed sensor shows the advantages these sensors afford. Called the Chariot of Fire, this sensor can detect changes in terrain, revealing possible locations of underground rocket launchers, a critical capability in a place like the Gaza Strip, where Hamas buries its rockets. Basically, the sensor can detect the invisible.

Israel's drones were originally designed for ISR missions—intelligence, surveillance and reconnaissance—to fly over targets and monitor developing situations. Early on, though, Israeli military planners understood that they could do more—that the unmanned aircraft could adapt.

The drones were already carrying laser designators, which could be used to "light up" targets that would then be attacked by helicopters or fighter jets. Why couldn't they carry the missiles too? Today, Israeli drones, including the Heron, reportedly have the ability to locate targets and destroy them as well.[7] Israel does not confirm that it has drones with attack capabilities. It is, however, well documented that this capability exists; Israeli drones have appeared at defense exhibitions with missiles mounted under their wings, and in WikiLeaks cables, Israel confirmed that some of its strikes in the Gaza Strip were carried out by armed drones. The Heron and the Hermes 450, another medium-sized drone developed by Israel's Elbit Systems, can reportedly carry laser-guided Hellfire missiles and smaller munitions like the

Israeli-developed Spike missile. The Spike causes less collateral damage and is said to be particularly effective in accurate strikes against wanted terrorists.

The Gaza Strip is ground zero for Israel's drone revolution. There, on a daily basis, the lawnmower hum of drones can be heard in the narrow alleyways. Gazans have given the drones the nickname "Zanana," Arabic for "buzz" or "nagging wife."

In Gaza, drones collect intelligence and help the IDF build its "target bank" in the event of a conflict. During Operation Pillar of Defense in Gaza, in November 2012, the IDF attacked nearly 1,000 underground rocket launchers and 200 tunnels that had been located and identified with intelligence gathered by drones. The first salvo of that operation was fired in a drone-assisted attack. Ahmed Jabari, Hamas's military commander, was driving in Gaza City when a missile struck his Kia sedan. Jabari, who had been at the top of Israel's most-wanted list and had escaped four previous assassination attempts, was finally taken out by a drone.[8]

Before Israel bombs Gaza in retaliation for rocket attacks, UAVs are there to survey the target; as helicopters and fighter jets move in to bomb a car carrying a Katyusha rocket cell, UAVs are there to ensure that children don't move into the kill zone; and when IDF ground troops surround a compound where Hamas terrorists are hiding, UAVs are there to provide real-time air support and guide the soldiers safely inside. And when needed, the drones can reportedly also attack.

At the smaller end of the IDF drone scale are drones not flown out of air force bases but pulled from soldiers' backpacks and literally thrown like a quarterback throws a football. One such drone, the Skylark, was delivered to IDF ground units in 2010. Weighing a mere 13 pounds, the Skylark fits into a soldier's backpack, but once airborne, it has an operational endurance of three hours at altitudes as high as 3,000 feet. These Skylarks can be utilized in all types of operations, from random patrols in the West Bank to large-scale ground offensives in places like Lebanon and Syria.

This new state of warfare provides commanders with quick over-the-hill intelligence. Commanders are no longer solely dependent on the IAF, which in turn can focus its attention on larger, more strategic missions.

The miniature UAVs are so popular that by 2016 they were being used by military forces in Australia, Canada, the US, South Korea, France, Sweden and Peru.

Drones are not only impacting the way wars are fought but are also changing the military's organizational structure, forcing all branches to adapt to the change that comes with new technology. After the establishment of the State of Israel, the Armored Corps' Seventh Brigade created a small, elite subunit called Palsar, a group of soldiers who functioned like forward sentries. They would advance before the tanks, scope out the territory and report back on enemy positions. In 2010, though, with the introduction of the Skylark, the IDF decided it was time to revisit the need for such a unit, or, at the very least, to redefine the soldiers' purpose. With drones able to fly ahead of the tanks, why put soldiers at risk?

✳ ✳ ✳

In 2009, Israel reportedly achieved a new level in drone performance.[9]

It was the middle of January, and Israeli soldiers were operating deep inside the Gaza Strip, the first large-scale ground operation since the "Disengagement," Israel's unilateral withdrawal from the Palestinian territory four years earlier. The government had just launched Operation Cast Lead in response to the firing of more than 2,000 rockets and mortars in the previous year alone. Prime Minister Ehud Olmert had decided: enough was enough.

While the country's focus was on the Israeli infantry and armored brigades operating in Gaza, a new threat was brewing far from Israel, in distant Sudan.

Intelligence obtained by the Mossad, Israel's super-secret spy agency, indicated that a ship packed with advanced Iranian weaponry—including Fajr artillery rockets—had docked in Port Sudan, on the Red Sea. These weren't ordinary rockets; they were game changers.

Up until then, Hamas's arsenal had enabled the Palestinian terror group to threaten the homes of the one million Israelis who lived in the south of the country. The Fajrs had the capability to go much farther and strike Tel Aviv. The containers, the Mossad learned, were being loaded onto trucks, to be transported north through Sudan and Egypt, where they would then be delivered to a depot near the Gaza border. Then the rockets would be smuggled into Gaza through underground tunnels.

The chief of staff of the IDF, Lieutenant General Gabi Ashkenazi, started drafting a plan to attack the convoy, but the clock was ticking. The moment the trucks crossed the border into Egypt, the strike option would be off the table. Israel couldn't mount an attack in Egypt, a country with which it had a fragile peace treaty. If the missiles then made it into Gaza, they would be swallowed up into one of the most densely populated territories in the world. While Israel's intelligence coverage over Gaza was good, it wasn't a sure bet. The rockets had to be stopped before reaching Gaza, meaning that the attack had to take place in Sudan.

An argument erupted within top defense circles. The doves— those opposed to the strike—warned of Israel's growing international isolation. The country was already under intense criticism for the rising death toll and extensive devastation in Gaza. News of a strike in another country would be difficult to explain. The hawks, on the other hand, argued that Israel could not sit by and allow advanced weaponry to reach Gaza. The potential threat was just too big.

The final decision was brought before Prime Minister Olmert, who had an affinity for covert, off-the-book missions. In 2007, despite US opposition, he reportedly approved Israel's strike against a nuclear reactor Syria was secretly constructing in the northwestern part of the country. He is also believed to have greenlighted a number of high-profile assassinations against top Hezbollah terrorists and Iranian scientists.

In operations like this, the prime minister usually asks a few technical questions about the mission and its risks before giving approval. In this case, in addition to the usual procedures, it would have been important to ensure that the strike could not be traced back to Israel. The mission would have to be done without leaving fingerprints.

The question now was how. Sending fighter jets to Sudan was risky. The entire mission could be jeopardized if there was a malfunction or one of the planes was detected by Egyptian or Saudi radars, which covered that part of the Red Sea.

There were also technical considerations, since the target—a convoy of trucks—would be on the move and, as a result, difficult to track. Timing was everything. The intelligence would have to be precise; the fighter jets wouldn't be able to stay in Sudanese airspace for very long, and they would have limited fuel.

The next step was for Ashkenazi to summon the head of his operations department and begin planning the mission. Together, they would need to consult with the air force's Operations Research Team, a group of engineers, scientists and munitions experts who evaluate targets and recommend the type of aircraft and bombs that need to be used to destroy them. Different options were considered, but the IDF reportedly chose an unconventional route—to strike the convoy with the help of drones.[10]

This was a first. While the Israel Air Force had reportedly used drones before for small-scale strikes—like targeting a lone terrorist in the Gaza Strip—they were mostly used for reconnaissance missions over places closer to home, like Gaza and Lebanon. They had never before been used in long-range strikes in a distant country like Sudan. Nevertheless, the decision made sense. Drones had the ability to linger for extended periods of time over an area like the vast Sudanese desert. There, they could hover and wait for the convoy to show up.

"When you attack a fixed target, especially a big one, you are better off using jet aircraft," an Israeli security source told the *Sunday Times* after the attack. "But with a moving target with no definite time for the move, UAVs are best, as they can hover extremely high and remain unseen until the target is on the move."[11]

As preparations moved ahead, the list of people involved was getting longer. So the IDF applied the strategy it often used when preparing for sensitive operations: compartmentalization. Only a handful of officers knew all aspects of the mission. Others were told only enough to keep them focused. Everyone knew that if word got out, the mission would be scrubbed, and the Iranian missiles would reach their destination in Gaza. The next time Israel saw them would be when they slammed into homes in Tel Aviv.

The yellow, sun-scorched Negev Desert is mostly barren, with little water or vegetation. Few Israelis settled there, leaving the large, dry terrain as the IDF's primary training ground. Israel's UAV operators were already experts at tracking moving vehicles, but until then, they had been focused on a single terrorist driving in a car or riding a motorcycle. To prepare for this mission, they had to practice locating and following a couple of trucks loaded with missiles. In the expansive Sudanese desert, that was like finding the proverbial needle in a haystack.

The Heron TP, Israel's largest drone, with the wingspan of a Boeing airliner, and the Hermes 450, the IDF's main attack drone, were the UAVs chosen for the operation. The Heron TP would fly in first, at altitudes where it could not be detected, to locate and track the convoy. The next wave would consist of Hermes drones and, if needed, fighter jets, which would dive in for the strike.

On the night of the bombing, there were some clouds, but for the most part the skies were clear, typical January weather in Sudan. As the Sudanese and Palestinian smugglers made their way through the desert, the last things on their minds were the Israeli drones tracking them from thousands of feet above. Even if they saw the incoming missiles, it would have been too late. Forty-three smugglers were killed, and all of the trucks were destroyed.

A Heron TP drone takes off in Israel. IAI

The initial mission was a success. A few weeks later, in February, Iran tried again. Olmert reportedly approved another strike. This time, 40 smugglers were killed, and a dozen trucks were destroyed.

The Sudanese were stunned. They had known that Iran and Hamas were using their country as a clandestine smuggling route, but Sudanese president Omar al-Bashir's government thought Israel would never do something as daring as launch an attack on a sovereign African nation. This analysis led the Sudanese government to the wrong conclusion: America must have been behind the strike. On February 24, a few days after the second strike, the chargé d'affaires at the US embassy in Khartoum, Alberto Fernandez, was summoned to the Sudanese Foreign Ministry, on the banks of the Blue Nile River, for a meeting with Ambassador Nasreddin Wali, head of the Americas Department.[12]

"I have sensitive and worrisome information to relate to you," Wali told Fernandez. The US official knew what was coming but played it cool. Looking down at his handwritten notes in Arabic, Wali pulled out a torn and worn-out map of Sudan and pointed at an empty patch of desert in the eastern part of the country. Fernandez listened as Wali read out the number of people killed and vehicles destroyed. "We assume the planes that attacked us are your planes," Wali told the American diplomat.

The suspicion was not baseless. A few weeks earlier, Fernandez had been at the foreign ministry to condemn the Sudanese government for allowing Iran to use its territory to smuggle weapons to Hamas in Gaza. Sudan assumed that after the official diplomatic protest, the US had decided to take military action.

Fernandez mostly listened as Wali lamented America's decision to unilaterally strike Sudanese territory and to undermine the two nations' "tight cooperation" on security.

"We protest this act and we condemn it. Sudan reserves the right to respond appropriately, at the right time, in a legal manner consistent with protecting its sovereignty," Wali concluded. Fernandez did not deny the Sudanese accusation but promised to relay the demarche to the State Department in Washington.[13]

Even if the US knew that the strike had been carried out by the IDF, as reported, Fernandez refrained from outing Israel to Khartoum. Nevertheless, Olmert could not hold back from publicly hinting at the possibility that Israel had been involved in the operation. A few days after Fernandez's meeting, the prime minister took the stage at a security conference near Tel Aviv and revealed that Israel had carried out counterterrorism operations in places "not that close" to home.

"We are hitting them, in a way that strengthens deterrence and the image of deterrence, which is sometimes no less important, for the State of Israel," the prime minister said. "There's no point getting into details, everyone can use his imagination. The fact is whoever needs to know, knows . . . there is no place where the State of Israel cannot act."

But while Israel maintained its silence, the operation shed light on the growing role that drones—of different shapes and sizes—are playing in Israel's wars. Today, drone flights account for nearly half of the overall annual flight hours in the entire Israeli Air Force, producing a few hundred hours of surveillance on a daily basis. Israel's infantry operations almost always include drone support. The air force top brass recently gathered for a special workshop called "2030" to strategize and determine how the force will look in the coming decades; at the end of the two-day seminar, most agreed that over the next few years, Israel will phase out its older F-15s and F-16s and very soon become a force consisting almost entirely of drones, some of them small enough to fit into your pocket.[14]

* * *

A visit to the Palmachim Air Force Base reveals another aspect of this story. Hidden among the rolling sand dunes are the headquarters of the IAF's first drone squadron. Young male and female operators dressed in green jumpsuits sit inside air-conditioned command caravans just a stone's throw from the Mediterranean's sandy shoreline watching surveillance footage stream in from the drones they are moving with black joysticks.

When Operation Cast Lead, Israel's anti-Hamas offensive, started in December 2008, Captain Gil (IAF personnel can be identified only by their first names), the squadron's deputy commander, was in the midst of planning a long-awaited vacation overseas. He canceled his plans and gathered his soldiers, the young men and women who serve as drone operators.

"We have two objectives," Gil told them. "The first is to provide support for our ground units. The second is to do everything possible to minimize civilian casualties."

One Sunday, Gil was on duty inside a command trailer. Working closely with Israel's various intelligence agencies, he had been tracking a man whom Israeli intelligence suspected was a senior Hamas terrorist. As the man left his home in a northern Gaza refugee camp, Gil followed from above.

Without knowing he was being watched, the man began doing something with what appeared to be wires. Gil's fear was that he was preparing a roadside bomb, possibly to be used in an ambush against IDF troops operating nearby.

"This looks suspicious," he said into his headset.

Back at IDF Southern Command, target specialists agreed with Gil's assessment and ordered a nearby aircraft to attack. But then Gil took a closer look. The man wasn't building a bomb with wires. He was hanging laundry. "Stop!" he yelled into the head-set. "Don't attack."

A few days later, Gil was again working his shift when he spot-ted what appeared to be an IDF soldier walking down a narrow alley in northern Gaza. But something was strange. Yes, the man was wearing an olive-colored uniform, but he was walking alone, and his M-16 rifle was slung over one shoulder and not across the chest, in violation of IDF regulations. "Why would a soldier be all by himself?" Gil asked another operator inside the trailer. Gil continued to track the man until he entered a home and emerged a few minutes later wearing different clothes. The man immedi-ately became a target.

On two different occasions, Gil experienced the advantage that drones provide on a battlefield—preventing an attack on an innocent man and ensuring that a wanted terrorist did not get away.

"By sitting far away, I have the ability to study a target for a long time and ensure that it is the right target," Gil told us when we met at the squadron headquarters. The responsibility is enor-mous, and Gil, who is in his thirties, is one of the older operators in the squadron. Most are in their early twenties. "What keeps me up at night is the responsibility that rests on us to do everything possible to minimize collateral damage," he said. "On the other hand, though, if I tell a soldier that an area is clear and he is hit by a sniper, then I am also responsible. There is always a risk."

Soldiers with so much responsibility at such a young age are forced to grow up quickly and think out of the box. UAV opera-tors and other soldiers in command positions in the IDF don't have the luxury of following a handbook with black and white operational plans and instructions. They have to think on their feet, make life-and-death decisions in split seconds and later that night go home to their families. As Gil put it: "Flying the drones? That's the easy part."

But what was the secret of Israel's success? How did such a small country develop a drone industry one step ahead of the rest of the world?

By constantly being on the front lines of war, Israel is forced to test new tactics and technology, sometimes decades before other Western countries. Since many of the needed platforms do not yet exist, Israel has to develop them on its own. That is what happened with drones.

Looking back at an aerospace engineering career spanning over five decades, Karem points to two factors that were key in turning Israel into the world's drone leader.

The first involves Israel's natural environment and surroundings. "I have worked with engineers and technicians in Israel, France, the United Kingdom, the US, Germany and Japan," he said. "Israelis don't have different genes than people from these countries, but what we do have is pressure—to win and to survive—and that forces us to be the best we can be."

The second advantage is the "fusion of thinking" that exists in Israel. All Israelis serve in the army. Those who then go to work in defense companies are still in touch with their army friends. They talk and share ideas. Everyone knows everyone. This helps shorten the development process and enables innovators to bounce ideas quickly off one another. If it's been tried already, they can find out and move on to something else. And if not, they can get real feedback from the battlefield, usually in just a single phone call.

Amit Wolff, a young aerospace engineer at IAI's drone division, is an example of this phenomenon. In the mid-1990s, after high school, Wolff was drafted into the IDF along with his fellow classmates. He was motivated like many other ideological young men and decided to try out for a relatively new elite military unit called Maglan, named after the Ibis bird. The unit had been established a few years earlier to operate deep behind enemy lines, with specially designed weapons systems. Like the bird, Maglan soldiers learn to blend into their surroundings, move covertly and locate their targets.

At the time of Wolff's army service, Maglan's focus was on the West Bank and southern Lebanon, where it conducted almost daily operations against Hezbollah terrorists. After his discharge from the military, Wolff studied engineering at the Technion and was hired by IAI to work in its drone division.

One day in 2008, he met with some of his colleagues at a coffee shop near the office for an informal brainstorming session. Since the days of his military service, he had pondered the possibility of designing a drone that could be easily carried on a soldier's back and quickly deployed to provide over-the-hill reconnaissance.

He remembered being frustrated while walking through dangerous and narrow alleyways in densely populated areas like the Gaza Strip and some parts of the West Bank, not knowing what was waiting just around the corner. "We started discussing the possibility of creating a lightweight drone, which could be easily unpacked, and take off and land vertically without the need for a runway," he told us when we met at IAI headquarters, outside Tel Aviv.

In the coffee shop, Wolff pulled a napkin from under his cappuccino and roughly sketched a design for a new drone. A few days later, he took his idea to the head of R&D, who listened to the proposal and decided to allocate an initial $30,000 for its development. The investment paid off, and two years later, IAI unveiled the Panther, one of the first tilt rotor drones in the world, which can take off and land like a helicopter and hover over targets at 10,000 feet for six hours at a time. The Panther is so light that a soldier can carry it on his back during operations.

"In general, our ideas come from a number of sources," the soft-spoken Wolff explained, as we walked the green lawns at IAI headquarters. "We follow inventions on the Internet, and we are in close contact with the defense establishment to understand the IDF's needs. But there's no question that a lot of us are inspired by our own experiences from the field, and since we were there, we know what is missing."

Wolff fused a number of identities. On the one hand, he was a soldier with extensive battlefield experience. But at the same time, he was also an aerospace engineer. That dual identity gave him the knowledge, experiences and skills needed to innovate and invent a new weapon system.

This intimate relationship between the civilian and military worlds is a national asset of immeasurable proportions.

As ironic as it sounds, another contribution to the development of Israeli drones was the cancellation of the Lavi, the most ambitious aircraft project undertaken by the government. Started in 1982, the Lavi (Hebrew for "young lion") was a multirole, single-engine, fourth-generation fighter jet developed by IAI as the future primary combat plane for the Israeli Air Force. Over a billion dollars were spent on its development—much of it in the form of US assistance—and a number of prototypes were manufactured and flown. But in 1987, the Israeli cabinet, in a 12-to-11 vote, decided to cancel the project, a decision that still stirs controversy among veteran defense officials. The program was canceled after the cost of the Lavi skyrocketed—so to speak—and amid immense pressure from the US, which wanted Israel to purchase the F-16 in its place.

While the Lavi cancellation was a big blow for IAI, which had to lay off hundreds of engineers, work on a combat jet provided Israel with institutional knowledge that it was able to apply to the country's other burgeoning projects—satellites, missile defense systems and drones. The talent focused on one single aircraft suddenly spread like wildfire throughout Israel's defense and high-tech industries.

❋ ❋ ❋

Israel's invention of drones has revolutionized the modern battlefield. It has allowed militaries to put fewer boots on the ground, to evaluate targets, to gather more accurate intelligence and to provide a level of superiority unmatched around the globe. This innovation was achieved in the most difficult of circumstances.

In the summer of 2006, though, that superiority came under question. On July 12, Hezbollah—the Iranian-backed terror group—invaded Israel and abducted two IDF reservists. Israel retaliated, and the monthlong Second Lebanon War erupted.

In the beginning, it didn't seem like a fair fight. Israel was a nation-state with the most advanced weapon systems in the world. Hezbollah was a terror group without bases, an air force or a navy. Its fighters wore civilian clothing and blended into their surroundings; its total budget amounted to a fraction of what Israel spent annually on defense. But just a few days into the war, Israel realized it had underestimated its enemy. Hezbollah not only fought well on the battlefield—121 Israeli soldiers were killed—but despite incessant Israeli airstrikes, it continued throughout the war to fire an average of 120 rockets a day. It hacked into the IDF's radio systems, launched sophisticated cyber attacks and cracked Israeli cellular phone networks along the border.

But the real surprise came toward the end of the war, when Hezbollah flew three Iranian drones into Israel, carrying a payload of over 20 pounds of explosives. The drones never reached their targets—two crashed and one was shot down by an Israeli jet. Nevertheless, their use sent shock waves through the IDF's top echelons. Israel, the state that invented drones, had become the first state to be attacked by drones operated by a nonstate actor, a terrorist organization. Israel's own weapon had been turned against it.

Another drone infiltrated deep into Israel in 2012, coming within just a few miles of the southern city of Dimona, home to Israel's nuclear reactor, before it was shot down. And drones were flown into Israel in 2014, this time by Hamas, from the Gaza Strip.

As in 2006, the drones were shot down by the IDF, but a video, released by Hamas, showed that these drones were carrying not just explosives but air-to-ground missiles—a technology not believed to be within Hamas's grasp and apparently installed on the drones for show. And in October 2015, Palestinian security forces announced that they had arrested members of a terrorist cell near the West Bank city of Hebron, who were plotting to launch an explosives-packed drone into Israel.

In these recent conflicts, Hezbollah and Hamas showed how warfare is evolving not just for Israel but for the entire Western world. While technology is an advantage, superiority cannot be taken for granted. Nonstate actors, like Hezbollah and Hamas, proved that they can beat states at their own sophisticated technology game.

The use of drones by Hezbollah and Hamas has forced the IDF to make certain adaptations, particularly within the air force. Until the use of drones by Hezbollah, the IDF assumed that it was the only operator of drones within the region and that even if Iran had a drone or two, they were a few generations behind Israeli drones and, anyhow, could not reach Israel. As a result,

Israeli radars were not designed or programmed to detect drones, only larger aircraft. The appearance of drones on its borders forced Israel to improve its detection capabilities—to invent new radars and retrofit older ones.

Israel has also reportedly taken the war against Hezbollah's drones into the shadows. In 2013, for example, Hassan Al-Lakkis, a top Hezbollah commander in charge of the organization's drone fleet, was shot dead in Beirut. According to Lebanese reports, two skilled assassins—disguised as tourists—approached Al-Lakkis's car and shot him with silenced handguns. Hezbollah accused Israel, which neither confirmed nor denied its involvement but had an obvious interest in undermining Hezbollah's drone capabilities.

While of great concern to the Israelis, the use of drones by Hamas and Hezbollah makes sense. Drones are small, are able to fly low and slowly, and are difficult to detect and track with conventional radar systems.

Rules of engagement are also foggy. If a drone invades a country, is that an act of war, like a fighter jet bombing or a missile strike? In 2012, when Hezbollah flew a drone near Dimona, Israel—to prevent setting off a larger conflict—decided not to respond. But what will happen when a drone strike against Israel succeeds? Will Israel use it as justification for an all-out war?

These questions have yet to be answered, and while drones, like robots, are not yet making strategic decisions on their own, they are, like never before, shaping the way wars are being fought.

3

ADAPTIVE ARMOR

"Start moving," Lieutenant Colonel Effie Defrin, commander of the 401st Armored Brigade's Ninth Battalion, yelled into the radio.

It was August 11, 2006, the beginning of what would become known as the Battle of the Saluki, a controversial last-ditch effort by the Israeli military to gain ground before a cease-fire went into effect and ended the Second Lebanon War against Hezbollah. The idea, hastily hatched at military headquarters in Tel Aviv, was to cross the Saluki River in southern Lebanon and conquer the territory believed to be used by Hezbollah for most of its rocket attacks against Israel. The expansion of the ground operation, Israel believed, would give it more leverage in the cease-fire talks at the United Nations. The war was coming to an end, but the government believed this last operation was worth the risk.

The Israeli tanks moved slowly along the narrow mountain path, vulnerable and exposed to anti-tank missile fire. The noisy and creaking tank tracks rolling over the rocky terrain put thoughts of Hezbollah squads out of the soldiers' minds. As it turned out, the IDF's prized tanks were rolling straight into an ambush.

Hezbollah reconnaissance teams identified the convoy of tanks as it approached the mountain crossing and immediately passed the information on to anti-tank squads waiting patiently in nearby villages. Because Defrin was the commander, his tank stood out with its numerous antennas. The Hezbollah fighters took up their positions and waited. They followed the customary procedure—identified the commander's tank, placed the Kornet missile target crosshairs on it and fired. Seconds passed. Suddenly, a muffled noise rocked the tank, and dust rose to the ceiling. Defrin kicked the gunner and turned to him angrily: "Are you out of your mind? You fired a shell?!"

"No way," the gunner stammered. "I didn't fire . . . I think we were hit by a missile."

The Merkava Mk-4 tank continued rolling. "There's no way we were hit by a missile . . ." Defrin mumbled to no one in particular. He then looked to the rear and saw three anti-tank missiles whizzing toward him. The static talking over the radio drowned out the buzz of the missiles. One hit the tank but did not penetrate. A second flew right over and missed. Defrin remembers only the noise from when the third missile struck. After that, everything turned black.

Israel hadn't been looking for war with Hezbollah, but on July 12, 2006, it was left with little choice. Hezbollah guerrillas crossed into Israel, attacked an IDF border patrol and abducted two reservists. Prime Minister Ehud Olmert, using the attack as justification to try to change the situation along Israel's northern border, led the country into its first war in more than 20 years.

A few months before the war, the IDF General Staff had met

for a two-day seminar to debate a series of proposed structural changes to the military; which, in the more immediate term, would include the closure of a number of units.

The forecast for the Armored Corps was grim. At the time, the IDF was focused on curbing Palestinian terrorism in the West Bank and the Gaza Strip. Tanks were perceived as irrelevant, and the General Staff was considering closing several armored brigades and reducing the number of tanks it manufactured annually. An hour after the Hezbollah abduction, another nail was driven into the Merkava coffin when a large explosive device detonated underneath a tank deployed along the Lebanese border. The tank crew was killed instantly, and the pride of the Israeli defense industry was shattered.

Defrin had originally planned to follow in the footsteps of his older brother, a paratrooper who had been seriously injured in a clash with Hezbollah guerillas several years before Defrin's draft date. The paratroopers were seen as the IDF's elite class. Members of the unit went on to fill top IDF brass positions, and several became chief of staff. In high school, Defrin jogged and lifted weights and passed the paratroopers' grueling two-day tryouts. But then the IDF doctors decided he was unfit for parachuting and while he asked to become an infantryman, they instead sent him to the less-prestigious Armored Corps.

In May 1991, Defrin was drafted into the Armored Corps' Seventh brigade and sent to basic training in the Arava Desert in southern Israel. The sandstorms and dry weather reflected his mood. He did not want to be there. At the entrance to the base, he saw a steel machine covered with a tarp that was supposed to conceal a state secret: Israel's new tank. Defrin could care less. The glimpse of the tank merely intensified his feeling of a missed opportunity. He continued to dream of running over hills with his face covered in camouflage and an M-16 slung over his shoulder. He filed several requests to move to the infantry corps, but they were all rejected. Months passed, and Defrin slowly came to terms with his sentence. By the end of advanced training, he had been selected as the unit valedictorian.

The unit's sergeant major used to yell at Defrin and his fellow soldiers regularly during roll calls. One day, as the soldiers stood outside in the Golan Heights rain and mud, soaked to the skin, the sergeant major screamed as he pointed toward the border. "There is Syria and here is Israel. You are in the middle. If anyone is protecting this country, it's you and the tanks. There is no one else," he said.

Defrin and his fellow soldiers got the message. On the other side of the border was the Syrian military, the last conventional Arab military with which Israel was still in a state of war. Surprise emergency drills, held day and night, were therefore part of the Israeli military's weekly routine. The commanders wanted to instill in the fighters the understanding that when Syrian divisions advanced toward the northern towns, every minute it took the fighters to get into their tanks counted. The question was when, not if, Syria was going to attack. The soldiers were always on alert.

It was with religious awe that the commanders spoke with the soldiers about the tank. On Fridays they would welcome in the Sabbath by preparing the tank, cleaning its inside and polishing it on the outside. The intimacy between the soldiers and their tanks was formed during combat training but also as the soldiers cleaned the tanks with buckets of soapy water and sponges. The tanks, they were told, had souls and needed to be cared for gently.

The early 1990s was a time of incessant fighting. Israel was bogged down in southern Lebanon, and when Defrin returned from officer training, he was sent on a series of raids against Hezbollah terrorist strongholds. There, for the first time, he encountered the tank's bitter enemy—the anti-tank missile. One night, Defrin was parked in his tank outside a Lebanese village; his mission was to provide cover for an infantry force on a nearby reconnaissance mission. He passed the time eating pita and hummus while his head jutted just a bit out the commander's hatch. Suddenly, a cloud of smoke flew right over his head. It had come from a Sagger anti-tank missile, which missed Defrin by just a few feet. Two weeks later, another missile was fired in the same sector. This time, though, the tank crew was not as lucky, and an IDF officer was killed.

At the time, the Merkava Mk-2 was the IDF's most modern and innovative tank, having entered service to replace the aging Magach, upgraded American Patton tanks that had been in use since the 1960s.

Now a company commander, Defrin was summoned one day in 2004 to the Nebi Musa training base in the Judean desert to see the new Merkava Mk-4 tank. Everyone knew that the IDF was working on a new tank, but none of his fellow officers had seen it. The tank was one of the most closely guarded state secrets at the time, and rumors were running amok about its revolutionary design and capabilities.

The officers were instructed not to take pictures of the tank and not to circulate any details of its performance. To many of them, it looked like a spaceship. It was larger than the tank they were used to operating and had a new and more powerful cannon than its predecessor. Its 1,500-horsepower diesel engine significantly enhanced the tank's speed, giving it the ability to cross complex terrains in record time. The sophisticated command-and-control systems gave tank commanders the ability to identify and fire at targets faster than ever before.

In 2000, the Second Intifada erupted, and the IDF was sent back into Palestinian cities in the West Bank. Budgets needed to be diverted to routine security patrols, the construction of checkpoints and security barriers. Among the IDF top brass, there was talk of restructuring the military, of abandoning the mandatory draft and creating a smaller, smarter and more professional army with significantly fewer tanks.

The number of tanks dropped to its lowest level since the Yom Kippur War. Training regimens were dramatically cut, and Armored Corps soldiers got to see their once-beloved machines only at ceremonies and in PowerPoint presentations projected in classrooms. Instead of defending Israel's borders from a Syrian invasion, the soldiers were sent on routine security patrols in the Gaza Strip, in the West Bank and along Israel's border with Egypt, where they looked for illegal immigrants and drug smugglers. Defrin began to forget what the inside of a tank looked like.

In the summer of 2006, all of that suddenly changed. The Second Lebanon War broke out, and Defrin's battalion was sent to the North for a short training drill to regain some of the basic skills needed to operate the Merkava. It took a few days, but soon enough Defrin and his soldiers regained their confidence and felt prepared to be sent into Lebanon. The tank was like a bike, some soldiers joked. You never forget how to ride.

For years, Defrin and the other tank crews had heard about the anti-tank missile arsenal that Hezbollah had accumulated since Israel's withdrawal from Lebanon in 2000 and that was supposedly waiting for them on the other side of the border. The Hezbollah arsenal was said to include some of the most sophisticated anti-tank missiles in the world—the Metis, the Fagot and the RPG-29, with its tandem explosive warhead. But one missile frightened them the most—the Kornet. The nightmare of Western armored corps, the Kornet was sold by Russia to the Syrian army and passed on secretly to Hezbollah as a personal gift from

President Bashar al-Assad. This laser-guided missile, one of the most dangerous and accurate in the world, came with a seven-kilogram tandem warhead, giving it the ability to penetrate up to 1,300 millimeters of armor. The missile is fire-and-forget, meaning that once locked onto a target, it will hit.

The operation on which Defrin was being sent—to the Saluki River—had been on the table since the war broke out, in mid-July, but was continuously modified and postponed until the night of Friday, August 11. Israel received word that the UN Security Council was going to meet that night to declare a cease-fire and end the war. After 34 days of fighting, Israel would have to comply. But Prime Minister Olmert wanted to try and sweeten the UN resolution, and get a better deal, with a more robust international force to monitor southern Lebanon. A last-minute Israeli push deep into Lebanon could do the trick.

Defrin didn't like the plan. It was already after midnight when the final orders came in to push all the way to the Saluki, some 10 miles from the Israeli border. This meant that his tanks would arrive in daylight and would be completely exposed to Hezbollah anti-tank squads. While soldiers from the Nachal and Golani infantry brigades were supposed to be helicoptered to the other side of the ravine to provide cover for the approaching tanks, Defrin and his men would still be exposed on a narrow pass they needed to cross to get to the other side of the Saluki.

The day before, Defrin, his company commanders and a few reserves officers sat around a sand tray simulating the operation and discussing its weak points. The reserves officers warned Defrin that he would be a sitting duck in the valley. "War is not an insurance plan," Defrin told them, predicting that one or two tanks, at most, would be hit. His prediction was based on IDF intelligence claiming that Hezbollah would have no more than two anti-tank missile squads in the area.

They were all wrong. After the first missile hit, Defrin managed to shout through the radio: "Commander here—do not stop under any circumstances . . ." His tank continued moving, but then the second missile flew by, and right after that, the third one hit. Defrin felt like he was suffocating, as if he had swallowed something too big for his throat. He blacked out.

"Commander is down. I repeat: commander is down," the operations officer in Defrin's tank yelled into the radio. No one knew Defrin's exact condition, but it didn't make a difference. There was no time to waste. The tanks had to keep moving. Hezbollah anti-tank squads were still out there with more missiles.

Defrin woke up spitting blood—a lot of blood. His lungs contracted, and once again he lost consciousness. The Kornet missile did not penetrate his Merkava Mk-4 tank, but the fight for Defrin's life had just started. An IDF doctor carried the battalion commander out into the open and started operating on him. From that moment, it was a race against time. Defrin was evacuated, under fire, back into Israeli territory to Ziv Hospital in Safed.

The battalion's operations officer pulled himself together, took command of Defrin's tank and pushed forward toward the Saluki. In the end, they got there with deadly results: 12 IDF soldiers were killed by missiles fired from nearly 20 Hezbollah anti-tank missile squads, and 11 tanks were hit. The IDF claimed that in the ensuing battle it killed dozens of Hezbollah guerillas. On the following day, the parties declared a cease-fire.

Defrin was hospitalized in intensive care for almost three weeks. The recovery was tough. Upon discharge, he returned to his battalion, debriefed his commanders and soldiers and then went to meet the bereaved parents, to look them in the eye and explain what had happened.

While Defrin slowly recovered, the Armored Corps had embarked on a new fight for survival. The Battle of the Saluki, with its casualties, wounded soldiers and damaged tanks, sent shock waves through the defense establishment. The arguments of 2004, supporting the downgrading of the Armored Corps, were once again heard in Defense Ministry corridors. The future of the tank was at stake. Budget cutbacks seemed inevitable.

A couple of weeks after returning to his base, Defrin was invited to the Merkava Tank Directorate, the branch of the Defense Ministry that oversees the design and production of Israel's tanks. While he had been confined to a hospital bed, tossing and turning in pain, his Merkava Mk-4 tank was undergoing meticulous inspections—each scratch was examined and X-rayed. Whole sections were disassembled and then put back together again. The military and Defense Ministry wanted to understand everything they could about the tank and the missile onslaught.

A senior officer in the unit placed a gray folder labeled "top secret" in front of Defrin and pulled out a photograph. It showed his tank, and the impact points of the two missiles that had hit it were marked with red arrows. Seeing the black and beaten tank for the first time was, for Defrin, like being thrown right back onto the battlefield.

"I see where I was standing and where the missile hit the tank . . . how is it possible that I'm still here? How come I'm not dead?" Defrin asked.

The senior officer explained that despite the multiple hits, not a single missile had penetrated his tank. The Merkava had withstood one of the most aggressive attacks known to date on a single tank. This Israeli machine had made history.

The picture provided all the convincing Defrin needed to get back behind the wheel. It was, as he later told us, confirmation that the Merkava is the epitome of "cutting-edge technology."

A few months later, the officer who had once dreamed of becoming a paratrooper was promoted to the rank of colonel. It would take Defrin a few years to open up about the Battle of the Saluki, but he would become one of the Merkava's strongest advocates.

But not everyone shared Defrin's faith in the tank. The media lashed out at the Armored Corps. "The turret is exposed," ran one headline in a leading Israeli daily.[1] British and American newspapers reported on the failure of the once-strong Merkava and questioned "how the vaunted tank became so exposed to Hezbollah rocket fire."[2]

A lobby was growing within the military to cut back tank production. "They are irrelevant," these officers claimed. The army, they argued, needed to invest in developing new, better-protected and faster armored personnel carriers. Anything but tanks.

The debate was bitter. Budgets were limited after the war, and a cut in tank production would actually open up scarce resources for other IDF necessities, like increasing training for the infantry, renovating bomb shelters, developing missile defense systems and more. The news coming out of Europe had a similar narrative. Western armies were reexamining the future of their tanks. The United States, as an example, was drafting plans to withdraw the tanks it had stationed in Europe, from bases where they had been since World War II.

Hoping to see part of the defense budget diverted to education, welfare and the health system, some politicians compared the Merkava to the Lavi, the fighter jet that had been developed by Israel in the 1980s and became the pride of Israel's defense industry but was later canceled by the government. That decision followed a fierce intragovernmental battle, after which the state decided to purchase combat aircraft from the US and invest its own money elsewhere.

The same, claimed some experts, should be done with the Merkava. The Ministry of Defense received proposals to consider alternative, reasonably priced tanks that could be purchased from the US and Europe. Another proposal called for moving part of the Merkava production line to the US. While this would cut costs, it would also increase the risk that some of the tank's secrets would leak out.

Defrin and his Armored Corps colleagues fought back. They knew that the tank was still relevant—that ultimately, during battle, only a tank could cover ground quickly enough to conquer territory. Yes, there were risks. But that didn't mean that the IDF should give up on the Merkava.

It took some convincing, but the defense minister and IDF chief of staff eventually agreed. They didn't close the Merkava program but also didn't simply keep things as they were. They did something even more interesting—they adapted.

*** *** ***

To this day, the Merkava tank is classified as one of Israel's most top-secret projects. For decades it has been kept under a veil of secrecy, so that on "doomsday" it can be thrown like an iron monster onto the battlefield and defeat Israel's enemies.

Many were involved in the development and production of the Merkava, but one IDF officer stands out: Major General Israel "Talik" Tal, the father of the Merkava tank.

Born in Israel in 1924, Talik learned early on what danger meant in the Land of Israel. It was 1929, and Arabs were rioting across the country. More than 100 Jews would be killed. One day, the doors to Talik's home in the northern city of Safed were sealed off by a mob, and the house was set on fire. It seemed like the end, until Talik's uncle ran up the street with a group of British policemen, who dispersed the mob. The uncle ran inside the house and rescued his five-year-old nephew. This near-death experience helped shape Talik's life.

He volunteered for the British Army at the age of 17 and fought as a tank gunner in World War II. After the war, he joined the Israeli underground and helped purchase weapons for the soon-to-be state. In the War of Independence, he served as commander of a machine-gun unit and quickly climbed the IDF ranks—serving as commander of the Armored Corps, head of the Operations Directorate and the Southern Command and, eventually, a special advisor to the defense minister. Talik passed away in 2010. A plaque with his name hangs on a wall in the Patton Museum of Cavalry and Armor in Kentucky, celebrating him as one of the five greatest armor commanders in modern history.

Israel's search for a tank started with the establishment of the state. During the War of Independence, for example, soldiers from the IDF's Seventh Brigade set off in unbearable heat to conquer Latrun, a former British police fort taken over by the Jordanian Legion. Despite numerous attempts, the IDF repeatedly failed to conquer the key site, which straddled the Jerusalem–Tel Aviv Highway. It simply lacked the means of penetrating the Jordanian fortifications.

Senior Israeli defense officials, politicians and lobbyists tried talking to Western countries about purchasing tanks for the IDF. Deals were reached, but threats of embargos were always in the air. Then came the Six Day War in 1967, during which Israel nearly doubled in size, conquering the Sinai Peninsula from Egypt, the West Bank from Jordan and the Golan Heights from Syria. Israel knew it was only a matter of time before its neighbors tried to reclaim their lost territory. If it was going to win again, it would need a stronger Armored Corps.

After the war, the IDF received its first batch of French and then American tanks. In the late 1960s, Israel bought the

Centurion, at the time the backbone of the British Army. Israel made a few modifications to the tank, installing an impressive 105-millimeter cannon and transforming its turret, giving it the name "Shot," Hebrew for "whip."

As part of the deal, Israel also received two Chieftains—Britain's top-secret tank, still under development and equipped with a 120-millimeter cannon. After a series of trials, Israel was ready to make a deal for more, but then the British backed away, citing political considerations.

The British decision startled Israel. The Soviet Union was continuing to arm Egypt and Syria. Israel needed new tanks but didn't have anywhere to buy them.

Cancellation of the deal left a deep impression on Talik. He understood that Israel had no one to rely on and came up with a revolutionary idea: Israel would build its own tank. Most people thought Talik was crazy. Until then, Israel had not built any of its primary military platforms—aircraft, navy ships or armored vehicles. But Talik insisted that it was possible. The study of the Chieftain had created some expertise in Israel, and Talik felt that there was a strong enough foundation to build on. He found a few partners and started creating a sketch of a tank. By 1969, the idea seemed viable. The question was whether it made financial sense and whether Israel really had the technology needed to develop a tank that could compete with the Soviet ones being supplied to Syria and Egypt.

In the summer of 1970, Defense Minister Moshe Dayan, the Israeli war hero, and Finance Minister Pinhas Sapir met to rule on Talik's tank idea. The meeting came after a team of security and economic experts had reviewed the proposal. All aspects of the project were studied: Was the tank Talik suggested even possible? and Would its development make economic sense for the fledgling state? For Sapir, the Merkava project had the potential to serve as a critically needed economic engine. The security benefit was secondary.

"I am for it," Sapir told Dayan. "Do you want it or not?"

Dayan was concerned that the financial investment would overshadow other military projects and limit procurement plans. But in the end, he agreed and gave the green light for the first stage: development.

* * *

The blaring ring from the phone startled Lieutenant Colonel Avigdor Kahalani. It was the spring of 1971, and on the phone was a woman who identified herself as "Talik's secretary." A driver, she said, would be coming in the morning to pick up Kahalani for a meeting. He should be ready early in dress uniform.

The next day, a dark-green Plymouth Valiant—the army-issued car at the time for senior officers—stopped at the curb outside Kahalani's home. The driver motioned for him to sit in the backseat. Kahalani did not have a clue what the meeting would be about, but it didn't really matter. Talik was a legend in Israel. If he calls, you come. The car stopped at the entrance to a big warehouse in the Tzrifin Army Base south of Tel Aviv. Kahalani got out of the car just as Talik appeared swinging open the large iron gate, startling a flock of pigeons resting on a nearby building. He motioned for Kahalani to follow him inside as he pulled back a camouflage net covering something in the center of the large hall.

At first, Kahalani wasn't sure what he was looking at, but after a few seconds it started to become clear. It was a tank, but not a regular one. This one was made of wood. The shape was strange as well. "The tank hasn't got an ass," Kahalani said. "Where's the engine?"

Talik explained the rationale behind the new tank while

walking in circles around his wooden creation. "It's a new design. Engine and transmission in the front and an exit hatch in the rear," he said. This was revolutionary. Until then, all tanks had their engines in the rear, and the entrance and exit from the tank was at the top of the turret, not at the back.

Talik had asked Kahalani to come see the tank so he could get the young officer's support. Kahalani was one of the IDF's up-and-coming armored commanders. He fought valiantly on his Centurion tank during the Six Day War and received the Medal of Distinguished Service. In 1973, during the Yom Kippur War, Kahalani would make history as commander of the 77th Battalion, when he succeeded in repelling the Syrian assault on the Golan Heights.

When the Yom Kippur War broke out, Kahalani was already on the Golan. He managed to pull together about 150 tanks from various units and led them into battle against a Syrian force nearly

A Merkava tank during a drill with IDF soldiers in northern Israel in 2015.   IDF

five times the size of his. After several days of intensive fighting, Kahalani succeeded in stopping the Syrian assault, destroying hundreds of enemy tanks and reoccupying the dominant positions Israel had initially lost on the Golan. For his actions, Kahalani received the Medal of Valor, Israel's highest military decoration. The meeting in 1971 was meant to reassure Talik that the tank he was building would be something young tankers— the likes of Kahalani—would want to fight in.

Even before the first kilogram of steel was poured, Talik envisioned the tank standing on the military parade ground ready for action. He didn't let the pessimists—mostly the Treasury officials who were concerned that their money was going to waste—get to him. Step-by-step, he obtained the funding, knowledge and connections needed to establish a production line that could cast the tank body, manufacture the cannon and develop optics and fire-and-control systems.

At the end of 1979, the first Merkava tanks were ready. Disagreements about upgrading the tank and correcting some remaining faults threatened to delay the project, but Talik's determination and charisma swept obstacles aside. Three years later, the tank demonstrated its operational capabilities on Israel's northern front during the First Lebanon War, and two years after, the second version—the Merkava Mark II—was already rolling off the production line. The Israeli tank had been born.

❋ ❋ ❋

"We have the Jewish genome," Talik used to tell his soldiers at the Merkava Tank Directorate. "This is what differentiates us from the rest of the world. However, this does not absolve us from learning. Only fools refuse to learn."

Talik liked to surround himself with brilliant engineers like

Yaron Livnat, a member of Israel's "armor elite," whose father had served as the head of the IDF tank maintenance unit. Livnat enlisted in the IDF as an academic and studied electronic engineering at the Technion. His dream was to invent new innovative missiles. But dreams only go so far, and after completing officer's training, Livnat was assigned to the Tank Directorate. He thought it couldn't get worse, but then Livnat was sent to join a real armor unit, out in the field.

Talik believed that technicians needed to be connected directly with the battlefield so they could understand the challenges soldiers faced and then come up with solutions that addressed real and not theoretical problems. Distance—whether cultural or physical—could not be allowed. Livnat was assigned to the Seventh Brigade for two months. He ran over hills while under fire, crawled in the sand, loaded tank shells and listened to the battle stories of his commanders and fellow soldiers. In his head, he was making a list of possible improvements that could later be implemented in the tank.

"No one understood what I was doing there. They thought there was something wrong with me. They told me it was idiotic to leave my office and go to the Golan Heights," Livnat recalled. "I fully understood what I had to see and feel in the field. Not to mention the operational ties that I established with the soldiers and officers who later became battalion and brigade commanders."

Talik liked Livnat since he saw that the young engineer had chutzpah, that he didn't always toe the line, that for him, rules were usually just recommendations. Talik saw a bit of himself in Livnat. One day, Talik invited the young engineer for a talk and asked him to serve as his chief of staff.

Despite the compliment, Livnat politely refused, but insisted on explaining: "I am a young engineer. The technological side fascinates me. I must stay in this world."

Talik wasn't used to being rejected, but he appreciated Livnat's candor.

"People like you, who tell me 'no'; it is usually the end of their career. In your case, you'll become my protégé," he told Livnat.

Talik's hard-soft demeanor enabled him to capture the IDF's best and brightest. He was also a man of his word, and from that day forward, Livnat received backing to develop breakthrough systems. Talik nurtured him, and Livnat was appointed head of Merkava Mk-3's Fire Control Project, a job that would ultimately win him the prestigious Israel Defense Prize.

The father of the Merkava project continued following global tank developments, and in the late 1980s he noticed that militaries were trying to come up with ways to make their tanks more accurate in motion. Talik understood that ground battles would not be the static engagements of World War II. The tanks needed the ability to shoot accurately at faraway targets not just while moving but while moving fast.

"You have to make the accuracy of shooting on the move like shooting when static. I want identical results," Talik told Livnat one day. "The systems must give the commander operational freedom. I don't want the commander to have to stop the tank in order to shoot and hit a target."

Livnat was startled by Talik's new request and dared to suggest that it was impossible. A tank in motion would not be able to hit targets as accurately as when standing still. "It's delusional . . ." he said. But Talik believed it was possible and forced the engineers to hunt for a solution. Their work culminated in the development of a new fire control system called "Baz," Hebrew for "falcon."

"Talik didn't have a strong tech side, but he had a sixth sense that wasn't based on science," Livnat told us. "Talik was a gunner in World War II and had a leather portfolio in his office with shooting tables, graphs used to track a tank's hits and misses. He fired so many times in World War II that it became a second nature to him."

What helped Livnat and his team find a solution was Talik's recommendation that they focus on shooting-error factors, the reasons why the tank couldn't stabilize the gun when in motion. Research revealed that the causes included the accuracy of the tank shells, the stability of the gun and the gunner's ability to identify targets clearly while on the move.

Livnat ran dozens of experiments and discovered that the existing fire control system knew how to correct itself and overcome changes but that the gunner didn't know how to take all of that into consideration. What this meant was that while the tank crew could measure the range, aim and shoot, the gunner couldn't stay focused and would miss.

In 1989, Livnat and his team made their breakthrough. They developed an automatic tracking system combined with a video camera, which relieved the gunner from having to calculate the range and direction and allowed him instead to focus simply on when to pull the trigger. Only when confident that he was locked onto a target would he launch the shell.

Talik always pushed the needle a bit farther than it appeared able to go. This way, by demanding what seemed impossible, he would at least get a result close to what he wanted.

At the same time, Talik also gave independence to his engineers. He believed in his people, knew that they were experts in their specific fields and that, often, they knew better than he did what needed to be done to improve the tank. Once, when Livnat came to him with 30 proposed tank modifications, Talik had him split the list in half based on importance. When Livnat came back an hour later with the new list, Talik, who was busy reading some other documents, didn't even look up. "Priority A is approved, priority B is not approved," he said, determining the future of the tank as if ordering a deli sandwich.

✳ ✳ ✳

From the beginning, Talik emphasized the need for the tank to be flexible and able to adapt constantly according to the changes Israel encountered on the battlefield.

For that reason, the tank underwent upgrades every few years until 2003, when the IDF deployed the Merkava Mk-4. A major improvement over earlier models, it could drive and shoot faster and, more importantly, it came with a new modular armor kit. This meant that the tank could be fitted with the armor it needed based on the specific mission it was heading into. An area known to have anti-tank missiles required heavy armor. An operation without the threat of anti-tank missiles meant less. This also allowed tank crews to replace damaged pieces of armor on the battlefield without having to bring the tank back to a repair shop in Israel.

The ability to adapt is a prominent characteristic in the IDF and stems from Israel's limited resources. Unlike the US or some European countries, Israel cannot afford to simply shut down and open up projects as warfare changes. Instead, it needs to be able to extend the lifespan of its aircraft, navy vessels and tanks beyond the norm while ensuring that they adapt and remain relevant on the modern battlefield.

One change, for example, is in the targets Israeli tanks find themselves up against. In the past, tanks attacked other tanks. In today's Middle East, Israel doesn't have any enemies with tanks. Syria's military is almost completely eroded after years of civil war, and Hamas and Hezbollah don't operate tanks. This means that for tanks to remain relevant they have to be able to engage targets like a Hamas rocket cell hiding on the third floor of an apartment building or a Hezbollah terror squad hunkering down in a schoolyard.

To meet these challenges, the IDF has developed new weapons—sometimes satellite guided—that provide tank crews with the ability to strike buildings, anti-tank squads and even aircraft accurately. One such innovative weapon is the Kalanit, which can explode midair over terrorists hiding behind cover, or, alternatively, breach concrete walls and detonate only once inside a building.

What makes the Kalanit unique is the tank crew's ability to choose two different modes for the way it wants the shell to detonate. On the one hand, it can be used like a traditional shell to target fortified structures or other vehicles and detonate on impact. On the other, it is useful for targeting terror squads, which cannot be effectively targeted by a standard tank shell. In this mode, the Kalanit can be programmed to stop midair, just over the terror squad, and then explode with six different charges, scattering thousands of deadly fragments.

Today's tank also comes with the IDF's advanced Tzayad ("hunter") battle management system, basically a computer screen in the tank, which soldiers can use to see the locations of friendly and hostile forces. If a new enemy position is detected, all a commander needs to do is insert the location on the digital map. The position will then be seen by all nearby IDF forces—tanks, artillery cannons and attack helicopters.

A new version of the Tzayad software enables the system to recommend the type of munition that should be used to attack a specific target as well as the route a commander should take when leading forces into a combat zone.

A soldier practices using the Tzayad digital army program.   ELBIT

Tzayad shortens the sensor-to-shooter cycle—the time it takes from when an enemy force is detected until the point when that force can be engaged. According to some estimates, the IDF has the cycle down to just a few minutes.

The biggest change, though, came in 2012, with the introduction of the Trophy, a system that could intercept anti-tank missiles fired at Merkava tanks. Until then, the world had heard of defense systems that could intercept ballistic missiles like the Arrow, but not one that could protect a single tank.

The idea was actually born in the 1970s, after the Yom Kippur War, during which IDF tanks suffered heavy losses at the hands of Egyptian anti-tank squads. One officer came up with an idea to install hollow explosive belts around the tank, which would detonate if struck by an incoming missile. This way the missile would explode outside the tank and fail to penetrate.

With the introduction of the Merkava tank a few years later, the idea was put on ice. The Merkava had unprecedented armor. It didn't need an expensive active-protection system like the explosive belt.

But then came the Second Lebanon War and the Battle of the Saluki. While Defrin's tank remained intact, the threat Hezbollah's anti-tank arsenal posed needed to be dealt with. Intelligence showed that in Gaza, Hamas was learning the lessons from the war in Lebanon and was accumulating its own stockpile of advanced anti-tank missiles. And Syria was reportedly purchasing hundreds of motorbikes and training special forces how to ride and fire anti-tank missiles at the same time. These small, slippery targets would be difficult for tanks to locate and engage.

The belt idea was dusted off and handed to Rafael, a government-owned leading missile developer. The result was Trophy, a defense system that uses a miniature radar system to detect incoming missiles and then fires off a cloud of countermeasures— metal pellets—to intercept them. Trophy's radar also interfaces seamlessly with the Tzayad battle management system. This means that Trophy can automatically provide the tank crew with the coordinates of the anti-tank squad that fired the missile so it can immediately be attacked.

In the summer of 2014, the IDF used Trophy for the first time in combat when Israel launched Operation Protective Edge against Hamas in the Gaza Strip. It was the most extensive use of Merkava tanks since the Second Lebanon War eight years earlier, when Effie Defrin almost died trying to reach the Saluki River. This time, though, the tanks were unstoppable. Dozens of anti-tank missiles were fired at Israeli tanks. Most missed and 20 were successfully intercepted by Trophy. Not a single tank was damaged.

Israel was once again changing modern warfare.

* * *

But why Israel? Why did Israel understand, over time, what other countries didn't—that tanks could adapt and remain relevant even on the modern battlefield?

Part of the answer can be found in an old army base outside Tel Aviv. Called Tel Hashomer, the British base was captured by Israel during the War of Independence and became home to a number of units, including what is known as Masha, a Hebrew acronym that stands for the 7100th Maintenance Center, the place where the Merkava tanks are assembled and repaired.

Brigadier General Baruch Mazliach, commander of the Merkava Tank Directorate, recalls how as a young engineer he would wait in the field—sometimes even crossing enemy lines—for the tanks to return from operations. The engineers would examine every detail and debrief each tank crew member, thirsty for knowledge that would help them come up with ways to improve the tank. The engineers didn't sit in air-conditioned offices and wait for the soldiers to come to them. They learned from Talik that the connection to the field was critical.

In his office, Mazliach keeps a brown file marked "Top Secret." Its contents tell a story of a tank that was ambushed by Hezbollah in southern Lebanon in 1994. At one point, more than a dozen different types of missiles were fired at the tank, including mortar shells, which scored direct hits. Witnesses of the combined strike assumed that the tank would evaporate. There was no way, they thought, that the crew could sustain such an assault and survive. Clouds of smoke covered the entire area. The tank crew was feared dead.

Mazliach pulled a dusty photograph of the tank from the folder. "Each of the circles is a missile hit," he told us. "This tank was mercilessly attacked from every direction, yet only one of the soldiers was killed. On the one hand, the outcome was fatal and harsh, but on the other hand, it proved how well-protected the Merkava really is."

For the engineers like Mazliach, the tanks are not built for hypothetical scenarios. Those engineers spend time in the field, developing close relations with the soldiers and officers who serve in the tanks; often, the engineers' own children are drafted into the Armored Corps. One engineer's son was killed in an attack on a Merkava Mk-3. "We are like a family," he explains. "That's why everyone works 300 percent with their entire heart and soul."

Being on the front lines of conflict since its inception, Israel often is the first Western country to face evolving and new threats, sometimes years before the rest of the world. The firing of Sagger missiles at Israeli tanks in the Yom Kippur War was the first real use of the advanced Soviet anti-tank weapon in war. The use of Kornet missiles by Hezbollah in 2006 marked one of the first times a terror organization had used tactics and characteristics that traditionally belonged to conventional militaries.

This leaves Israel with little choice but to constantly adapt to changing reality and to develop weapons, like Trophy, that can be applied as necessary. Israel doesn't have the luxury to wait for these weapons to be developed somewhere else. It needs its tanks to fight, and it needs them to be protected.

That is why, in 2012, the IDF established a technical team to begin designing its future tank—aptly named "Rakia," Hebrew for "heaven." The significant expected changes will be in its mobility, crew size and fire-and-control systems.

But despite Israel's continued technological developments, in January 2015, the IDF received a painful and stark reminder of the advanced arms that are circulating throughout the region. Hezbollah fired five Kornet missiles at an Israeli military convoy patrolling the border with Lebanon. Two soldiers were killed and seven were injured. Similar anti-tank squads are believed to be scattered throughout the nearly 200 villages in southern Lebanon, waiting for a future Israeli invasion. The guerillas are dressed in civilian clothing, living in regular homes. Attacks in a future war could come from anywhere—schools, hospitals and even ambulances.

In the Middle East, warfare is constantly changing. In Gaza in 2014, the IDF watched as Hamas terrorists jumped out of cross-border tunnels in surprise attacks; ISIS fighters in Syria drive commercial vans when attacking villages, and radical Salafi groups in the Sinai have succeeded in seizing armored personnel carriers for attacks on Egyptian military posts.

The key word for Israel remains "adaptation." With the winds of war blowing along Israel's borders, the next test of the Israeli Merkava is only a matter of time.

4

CHUTZPADIK SATELLITES

Five years after the devastating Yom Kippur War, Israel's diplomats were bracing for their first attempt at peace with an Arab country. Not just any Arab country, but the biggest—Egypt.

It was the spring of 1978. A few months earlier, Egyptian president Anwar Sadat had made his historic visit to Jerusalem, where, in a speech at the Knesset—the Israeli parliament—he called for an end to 30 years of war and bloodshed. Secret diplomatic cables were flying daily to and from Washington, Jerusalem and Cairo ahead of the Camp David peace talks to be held that September.

But Israeli colonel Haim Eshed was not comfortable waiting. The stakes were too high. Eshed knew that a peace deal would need to entail a complete Israeli withdrawal from the Sinai, the massive piece of land Israel had conquered from Egypt during the Six Day War in 1967.

The problem was that Israel needed the Sinai. The territory served as a buffer between pre-1967 Israel and Egypt. If Egypt launched another surprise ground invasion, it would have to first reconquer the Sinai, and Israel would have time to prepare. Israel couldn't withdraw before finding a way to keep eyes on the ground.

As the head of the research and development division in Israel's Military Intelligence Directorate—Aman—it was Eshed's job to come up with technological solutions for operational problems.

He saw a solution to the upcoming withdrawal from Egypt: satellites. Not just any satellites though; blue and white ones. Made in Israel.

Eshed was something of an anomaly within the military. While IDF soldiers were known for being brash and undisciplined, Eshed was courteous and polished. He spoke impeccable English.

Born in Istanbul in 1939, he immigrated to British-controlled Palestine a year later with his family. When Israel was founded, he was just eight years old, but the country's resilience and fight for survival inspired him to make a career of the military.

With a knack for technology, Eshed decided on a multidisciplinary education. He received an electrical engineering degree from the Technion and a computer science degree from UC Santa Clara, in California.

When he enlisted in the IDF, his strong technological background was quickly put to work, and Eshed, in turn, quickly climbed the ranks. He was a legend in military circles after receiving a Medal of Honor a few months before the Six Day War, when he was just a low-level corporal.

To this day, all that can be said about the reason he received the medal are the few words written on the certificate by then–IDF chief of staff Yitzhak Rabin, who would go on to serve as Israel's defense minister and prime minister: "Haim Eshed demonstrated superior technical skills and made a huge contribution to the IDF's state of readiness."

Persuading the government to invest in satellites, though, was a long shot, even for someone with Eshed's track record. Space was supposed to be off-limits to small countries like Israel. It was the sole domain of the superpowers. At the time, only seven countries had launched satellites into space. The last to do so had been the United Kingdom, far back in 1971.

The idea was bold for another reason. Until then, Israel's space experience was limited, extremely limited. In 1961, it launched a meteorological research rocket called Shavit (Hebrew for "comet"). Research, though, was just the excuse. Israel's real motivation was to launch a surface-to-surface rocket before Egypt, which the Mossad believed was working on developing a rocket with help from German scientists.

In 1965, Israel considered establishing a real space program. The proposal, brought to the government, called for an investment in original space research as well as the development of satellites and rockets for civil use and possible military applications. The proposal was ahead of its time and was rejected.[1]

Eshed knew all of this, but he was determined. He first sought approval for the idea from his boss, the commander of Aman, Major General Yehoshua Saguy. He was hooked and gave Eshed the green light to move ahead. Together, they first shopped the idea around the air force. The pilots there weren't interested. "It's beyond our technological means," they told Saguy and Eshed.

They then brought the idea to the IDF chief of staff, Lieutenant General Raful Eitan, who dismissed it as "Luftgesheft," a Yiddish expression that translates literally as "air business" and is used to describe something that is a complete waste of time.

Israel, Eitan argued, needed to begin focusing on the withdrawal of military bases and towns from the Sinai and their relocation inside the country. A satellite would be a waste of time and money. Anyhow, the chief of staff pointed out, the air force had told him that its aircraft could provide all of the aerial reconnaissance the country needed even after a withdrawal.

Eitan also claimed that once there was peace, Saguy could simply drive into the Sinai on his own to look around and see if the Egyptians were preparing for war. "That's impossible," Saguy responded. "I would need to look inside every single bedouin tent to really know what is going on."

Eitan's opposition to the satellite project had support in the highest echelons of power. A little over a year earlier, Prime Minister Rabin had gone to Washington for meetings with President Gerald Ford. Rabin was given the royal treatment as the first head of state to visit America during its bicentennial year and was asked by Speaker Carl Albert to address a joint session of Congress.

But in a meeting with some congressmen, Rabin was taken by surprise when asked why Israel needed a satellite. Apparently, the Defense Ministry had submitted something of a "wish list" to the US with a variety of military platforms and weapons it wanted to buy, which included a $1 billion satellite. The congressmen were concerned that such a sale would hurt prospects for peace in the Middle East.

In his 1979 biography, Rabin described the episode: "I was pushed to the wall with embarrassing questions about procurement lists we handed out to the United States. The question, 'Why must Israel have . . . satellites, costing a billion dollars?' had no other answer than the serious and obvious one: 'We do not need such a system.'"[2]

The problem was that reconnaissance flights over Egypt—the alternative to satellites—were conducted in a way that ruined the quality of the photos. Rabin became prime minister after the Yom Kippur War, during which more than 100 Israeli combat planes were shot down, mostly by Soviet surface-to-air missile (SAM) systems. Fearing that reconnaissance missions over Egypt would spark a new military crisis, Rabin ordered the IDF to approve each flight with him beforehand. As a result, flights over Egypt were a rarity, and instead, the IAF had to take photos while flying within Israeli airspace and on an angle. The photos were simply not of the quality needed to really know what was going on over the border.

Then there were the rumors. Senior officers in Aman were apparently trying to get Eshed fired. One of the officers had even gone straight to Saguy, urging him to get rid of Eshed, while warning that the satellite project would ultimately fail after eating up crucial budgets.

To those who asked, Eshed explained that he wasn't interested in the rank or promotion that would come with the success of such a project. It was something far more basic—the need to ensure Israel's national security.

Not getting anywhere through standard IDF channels, Eshed decided to try his luck with Ezer Weizman, the decorated fighter pilot and, at the time, Israel's minister of defense.

Weizman might have been way up the chain of command and out of reach for a regular IDF colonel, but Eshed knew him from a joint operation they had worked on in the 1960s. Still, he kept news of the meeting quiet.

That's how, one humid summer day, Eshed walked inside the Defense Ministry building in Tel Aviv. After showing his ID to the security guard at the entrance, he took the stairs to the second floor and turned left down the hall to the defense minister's office. The chief of staff's office was on the opposite end.

This was the floor where the tough decisions were made. It was here where the IDF plotted the surprise attack on the Syrian and Egyptian air forces as the opening to the Six Day War and where the IDF brass convened on the holy day of Yom Kippur in 1973 after the country came under surprise attack on both its southern and northern fronts.

As he sat waiting for his appointment, he looked up at the photos of the former defense ministers hanging on a nearby wall. It was a mix of politicians, including David Ben-Gurion, Israel's founding father and creator of the modern IDF, and Moshe Dayan, the celebrated war hero who led Israel to victory in the war of 1967 but then to disaster in 1973.

Weizman came from Israel's elite. His uncle, Chaim, was Israel's first president. Born in Tel Aviv, Weizman enlisted in the British Royal Air Force during World War II and became a fighter pilot. He later helped establish the Israeli Air Force and build it into a force with a clear qualitative edge over Israel's neighbors.

When the secretary called Eshed, he stood up, straightened his uniform one more time and walked into Weizman's office. Weizman sat behind a large wooden desk. He usually wore a white shirt with two buttons open, revealing a bit of chest hair, gray like the color of his military mustache.

Behind Weizman, hanging on the wall, was a large satellite map of Israel and the surrounding Arab countries as well as a couple of photos from his days as a fighter pilot. On a side table was the special phone the defense minister had recently installed so he could communicate directly with Sadat's office in Cairo.

Eshed began his spiel.

"We have a problem and I have a solution," he told Weizman.

If before the withdrawal the Egyptians decided to attack Israel, he explained, they would first have to cross through the Sinai, giving the IDF plenty of time to mobilize troops to stop an advance on Israel proper. But now, after the withdrawal, Israel wouldn't know if they were even there. It needed eyes on the ground.

Eshed then pointed at the long border between Israel and Egypt on the map behind the defense minister. "Satellites can help us keep track of what is happening there after we pull out," he said.

He then dismissed the air force's suggestion that it could just fly spy planes over Egypt when needed.

"We can't do that to a country we just made peace with. That would be a violation of the treaty," he told the defense minister. "Satellites, though, can do it for us."

Weizman asked a few technical questions. He was interested in how much satellites would cost, how long it would take to develop them and whether the quality of the photos would be good enough for intelligence purposes.

When he left the defense minister's office, Eshed didn't have what he had come for but had received what he needed—a chance, or, as Weizman sarcastically put it, "a chance to fail."

Within weeks, Eshed was on an El Al commercial flight to the United States to visit NASA's Goddard Space Flight Center in Maryland, home to the development and production of some of America's most advanced satellites. He then flew to France and visited the European Space Agency.

Eshed's conclusion was that Israel could do this.

The problem now was funding—not yet for building a satellite but for a more basic stage, the commissioning of a feasibility study from Israeli defense firms.

But persistence will get you only so far. Sometimes luck gets you the rest of the way.

In early 1981, Saguy traveled to Langley, Virginia, for talks with the CIA. While Israel had signed a peace treaty with Egypt and was gearing up for its withdrawal from the Sinai, Saguy had come on other business. The IDF was planning a military strike against a nuclear reactor Saddam Hussein was building in Iraq, and it needed satellite imagery.

As the first head of Aman to be promoted from within its ranks, Saguy was a master intelligence officer with a keen attraction to technology. During his tenure, Aman made some impressive technological leaps, including the invention and use of frequency-hopping transmitters for agents sent on covert missions abroad. Jumping frequencies made the transmitters almost impossible to intercept and record.

Saguy especially liked Visint, the acronym for visual intelligence, such as that from satellites. While phones can be blocked and signals disrupted, there was no technology at the time that could change what was caught on camera. The only downside, of course, was that a picture alone could not usually determine the intentions behind what was being photographed. If an enemy armored division is seen mobilizing, for example, additional intelligence is needed to know if it is doing so for an exercise or for war.

Saguy's request, though, was not a simple one. Israel's relationship with the CIA, when it came to satellite imagery, had known its ups and downs. At times, the relationship seemed to depend on who was director of the CIA.[3]

At the beginning of October 1973, for example, Israeli military attachés went to the Pentagon and requested information, based on US satellite imagery, about the deployment of Syrian and Egyptian forces. Israel had intelligence indicating that the countries were preparing an invasion but wanted to back it up with real proof. The attachés were told that the US satellites were not working and that photos were, therefore, out of the question.

When George H. W. Bush became director, in 1976, the attitude changed, and he agreed to provide Israel with actual imagery.

Aman jumped at the opportunity and sent two analysts—one
of them had started his career in the IDF as a chef—to Langley.
They sat in a separate room near the CIA's satellite analysts. There,
they could make specific requests, receive photos, analyze them
and then relay the information back to Tel Aviv.

Bush's successor, Stansfield Turner, overturned that policy
and agreed only to give the Israelis information gleaned from the
satellite's reconnaissance missions, not actual imagery. In 1981,
the policy was again reversed with the appointment of William
Casey. He did, however, restrict the photos provided to Israel to
those of targets that posed a potential and direct threat to Israel's
security.[4]

When Saguy arrived in Langley, he asked not only for imagery but also for direct access to a US reconnaissance satellite. This
was a huge ask, but Israel had prepared its case: the US was planning to sell an advanced airborne warning and control system
(AWACS) aircraft—basically a flying radar station—to Saudi Arabia. If the US was committed to retaining Israel's qualitative military edge, the Jewish State deserved something by way of
compensation. So it gave the US two options: either provide it
with "full and equal access" to an existing satellite or give Israel
exclusive use of a new satellite and ground station.[5]

While the US said it would consider the request, it rejected
Saguy's smaller request for images of the Iraqi reactor. When he
returned to Tel Aviv, frustrated, a few days later, he did something
unusual. Instead of going back to the General Staff and again arguing for funds, he simply rerouted $5 million from his own bud-
get at Aman and summoned Eshed.

"I'm green-lighting the satellite study," Saguy told Eshed.
"Don't let me down."

Eshed contacted two top Israeli defense contractors to request
proposals. Usually such requests need to be approved by the De-

fense Ministry, but Aman, as an intelligence agency, often by-passed bureaucracy, citing "national security."

The program had two main hurdles to clear. The first was to see if Israel could in fact develop a satellite with an electro-optic camera capable of taking photos from space at a high enough resolution that they would be of any value. Aman set a five-foot standard for the cameras, meaning that they would need to be able to distinguish between a tank and a truck.[6]

The second hurdle was to develop a launcher.

Eshed set up a meeting with an engineer named Dov Raviv, the Romanian-born head of the missile factory at Israel Aerospace Industries. Known by its Hebrew acronym, Malam, the factory was where Israel is believed to have developed the Jericho, a long-range, three-stage solid fuel ballistic missile that can reportedly carry a nuclear warhead and strike in the heart of any Arab capital in the Middle East.

What Eshed needed to check was whether IAI could build a missile that could be used as satellite launcher. Eshed had unique requirements, such as the direction Israel would need to launch the satellite. Until then, all countries launched satellites to the east, in sync with the earth's rotation. Israel, however, couldn't launch its satellite to the East, in the direction of Jordan and Iraq. If the launcher accidentally landed in an Arab country, it could be viewed as an act of aggression and lead to war. In addition, if the satellite crashed, Israel would have dropped a technological treasure chest into the hands of its enemies.

As a result, Israel needed to launch its satellites to the west, against the earth's rotation. Simply put, this meant that Israeli engineers needed to develop an extremely powerful missile launcher, since the missile would be flying not just against gravity but also against the earth's orbit when trying to place a satellite in space. Despite the challenge, Raviv said it was possible.

✳ ✳ ✳

On June 7, 1981, eight Israeli F-16s took off from an airfield in the Sinai and flew to Iraq, where they successfully bombed the Osirak reactor. The strike caught the world by surprise. While Israel's concerns about Iraq's nuclear program were well known, no one thought Israel had the operational capability to carry out a long-range strike of over 1,000 miles—to Iraq and back.

The Israeli jets flew through Saudi and Jordanian airspace en route to the Iraqi reactor. At one point, when flying over the Gulf of Aqaba, the jets were spotted by Jordanian King Hussein, who was vacationing there on his yacht. He noticed the Israeli markings on the aircraft and called the military to immediately send a warning to Iraq. The warning never made it, and the IAF jets sneaked into Iraq undetected.

The world was shocked and condemnations were quick to come, including from the White House. The F-16s used in the strike had only recently arrived in Israel. The planes were originally sold to Iran, but after the Islamic revolution, in 1979, the US offered them to Israel. Under pressure to punish Israel, President Ronald Reagan suspended the delivery of additional combat aircraft to the state, and the CIA kicked the Israeli analyst team out of Langley.

The US also launched an investigation to determine how Israel had obtained the necessary targeting information, even after their request for imagery had been rejected. CIA deputy director Bobby Inman ordered an immediate review of all the images Israel had requested and been provided in the previous six months.

While Israel, under Casey's policy, was supposed to receive imagery only of potentially direct threats, Inman discovered that it had in fact received images of areas not just in Iraq but also in Libya, Pakistan and other countries far from the Jewish State.

Furious, Inman set new criteria restricting the transfer of photos to Israel to those of areas within a 250-mile radius of the country's borders. Requests for farther-away targets could still be made but, under the new guidelines, would need to be approved by the CIA director himself on a case-by-case basis.

Protests by Ariel Sharon, Israel's defense minister at the time, to Defense Secretary Caspar Weinberger didn't help, and Inman's new restrictions remained in place.

Back in Israel, Eshed had completed the feasibility study and reached the conclusion that Israel had all of the necessary technology and technical know-how to build its own satellite. With the new CIA restrictions on satellite imagery in place, the officials who had initially opposed Eshed's plan were beginning to come around. Even they realized that Israel needed an independent capability. Dependence on the US was undermining Israel's national security.

"If you are fed from the crumbs of others according to their whim, this is very inconvenient and very difficult," explained Meir Amit, a former director of Aman and the Mossad. "If you have your own independent capability, you climb one level higher."[7]

In the US, the controversial strike in Iraq brought Israel's request for direct access to a satellite back on the table. Those in favor of granting the request argued that if Israel had had satellite access, it might have refrained from attacking Iraq, since it would have had a better handle on what was really happening at the nuclear site. Those against granting the request warned that access to a satellite would give Israel the ability to plan additional strikes throughout the region. The opponents also feared that if the US granted Israel access to one of its satellites, the Soviets would do the same for the Arabs.

A few weeks after the successful strike against the Iraqi reactor, Prime Minister Menachem Begin convened a meeting to discuss the satellite proposal. Begin would either approve the plan and allocate the necessary budget or bury the dream of an Israeli satellite forever.

It was a gamble. Saguy had met before with Begin. As head of Aman, he was responsible for Israel's national intelligence estimates and often met privately with the prime minister. During one such meeting, Saguy raised the satellite idea. Begin listened but voiced caution. First, there was concern that Israel would fail, the world would find out and the country's reputation for deterrence would be shattered. Second, there was concern that the world would view an Israeli satellite as a threat, or too much power for the small Jewish State.

"Make do with what you have," Begin told Saguy at the time.

Walking into the meeting, Saguy and Eshed knew that this was their chance. At the very least, they believed that the fate of the project was in the best of hands. They had great respect for Begin and knew that if there was an Israeli leader who could understand the strategic significance of developing an independent satellite capability, Begin was that leader.

Born in Lithuania, Begin became enthralled at a young age with Zionism and the dream of an independent Jewish state. Anti-Semitism and the persecution of Jews featured prominently in his life; his father once came home having been beaten badly by a Polish policeman whom he had tried to stop from cutting off a rabbi's beard. Begin carried that memory with him throughout his career.

"I have forever remembered those two things from my youth: the persecution of our helpless Jews and the courage of my father in defending their honor," Begin told President Jimmy Carter in one of their first meetings.[8]

Begin may have been a small and pale child, but like his father, he fought back against the Jew-haters he met on the school playground. Nineteen-thirty was a transformational year for the teenage Begin. Vladimir Jabotinsky, a Russian author who had founded the Revisionist Movement, was coming to Brisk to speak. Jabotinsky was pushing a revised version of Zionism. He believed that a Jewish state needed to be established on all of the Land of Israel without compromise. The 17-year-old Begin snuck into the theater where Jabotinsky was speaking. By the end of the speech he understood that only a state could protect and defend his people. There was no future for Jews in the Diaspora.

Begin joined Betar—the Revisionist youth movement—and quickly climbed its ranks. When World War II broke out, he joined the Free Polish Army and was sent to Palestine. A year later, the Nazis arrived in his hometown, rounded up a group of 500 Jews, including his father, and drowned them in a nearby river. His mother was later pulled out of her hospital bed and murdered.

By 1943, the 30-year-old Begin had been appointed head of the Irgun, an underground Zionist paramilitary group that had broken off from the Haganah, the Jewish community's main paramilitary organization. At the time, the Irgun was falling apart. It had only a handful of followers, even fewer weapons and no clear direction. Begin put the organization back on track and embarked on a series of pinpoint strikes against the British, culminating in the 1946 bombing of the King David Hotel—home of the British Administration—which killed 91 people.

The attack was devastating, but for Begin it represented a simple equation: the British needed to withdraw from Palestine for the State of Israel to be established, meaning that even a deadly attack like this was legitimate.

In 1977, Begin came to power as head of the right-wing Likud

Party, ending almost 30 years of left-wing rule. A new era began for the country, and Begin's government adopted policies very different from those of the previous Labor-led governments.

Throughout his career as an underground fighter and politician, the Holocaust cast a large shadow over Begin. It is often cited as the primary motivation that brought him to Camp David in 1978 to negotiate a peace deal with Egypt. In deliberations ahead of the bombing of the Iraqi reactor in 1981, Begin often dismissed criticism of the operation, saying: "I will not be the man in whose time there will be a second Holocaust."[9]

He later revealed that when the F-16s were in the air on the way to their target, his thoughts wandered to the Holocaust and his parents.

Back at the meeting, Eshed presented the satellite idea, and Saguy voiced his support, backing up the claim that Israel could no longer rely on allies like the US. It needed independence.

Then the debate began. Participants like IDF chief of staff Eitan expressed skepticism, arguing that an Israeli satellite program could turn into a major waste of resources and time, neither of which the military had to spare. Other officers tried to persuade Begin to invest in developing a guided cruise missile instead of wasting money on a satellite launcher.

But Begin was intrigued. While he avoided details—technology was definitely not his strong point—the prime minister asked a few questions to ascertain whether the satellites would play a critical role in Israel's defense rather than merely being used for commercial purposes.

Begin liked what he heard: an independent Israeli satellite capability fit into his overarching belief that less than 40 years after the Holocaust Israel could not deposit its fate in the hands of others.

That same ideology led him to authorize the attack on the

Iraqi reactor in spite of fierce US and European opposition. If the Holocaust taught Begin anything, it was that Jews should never rely on anyone, even good friends, when it came to something as basic as ensuring their survival.

"This will be the realization of the Jewish genius, which has the ability to do wonderful things," Begin told Eshed. "Get going."[10]

Eshed left the room while Begin and the other participants moved on to discuss additional issues on the day's agenda. What they couldn't have known then was just how monumental a step they had just taken and what it would do to secure Israel's future as a military superpower.

But it wasn't easy. The first hurdle was how Israel would gather information on building a satellite without revealing that it was doing so.

Begin knew the Americans would not be happy if they discovered Israel was building its own satellite, which they had already warned could set off an arms race in the Middle East.

So he came up with a crafty idea: to create a civilian organization—the Israel Space Agency—and to appoint as its head a man by the name of Yuval Ne'eman.

A former science minister in Begin's government, Ne'eman was a renowned theoretical physicist who brought military and academic clout to the position. He was a former senior intelligence officer in Aman and later served on Israel's Atomic Energy Commission. When President John F. Kennedy quizzed Israeli prime minister David Ben-Gurion about Israel's nuclear program in the 1960s, it was Ne'eman who provided the answers.

For the plan to work, Begin decided that only Ne'eman would speak publicly in the name of the Israel Space Agency. His name would appear on publications; everything would be done under the guise of scientific research. Eshed would appear on the board, but only with his academic title of professor. There would be no mention of his military rank and position.

Eshed and Ne'eman immediately understood that their satellites would need to be of limited size, mostly due to the limitations of the launcher, which would not be able to carry a heavy satellite. As a result, the first satellite, called Ofek (Hebrew for "horizon"), was designed without a camera. The idea was to first see if the Shavit launcher could place the satellite in space.

The satellite needed to be not just small, but really small. While America's main workhorse at the time—called KH-11—weighed over 13 tons, Eshed and IAI were designing a satellite that would weigh a mere 155 kilograms.

"Every kilogram counted," Eshed recalled. "Everything needed to be made special for this project."

The next challenge was to find funding. While Begin had approved the program, the skepticism in the IDF remained, and the General Staff was reluctant to allocate the necessary funds, which some estimates put at nearly $250 million, just as an initial investment.

But Saguy had a solution. In his previous position as deputy head of Aman, he had forged a close relationship with a number of senior South African officers, some of whom would travel to Israel with their wives. Saguy and his wife, Hanna, would host them at their large home, in a small pastoral town south of Tel Aviv.

Israeli–South African ties were in their heyday in the 1970s. South Africa needed weapons, and Israel needed money. Saguy raised the idea one day with one of his counterparts and succeeded in getting a commitment of several hundred million dollars for the project. The agreement was classified as "top secret" and kept that way for about 15 years, until South Africa revealed that it had jointly funded an Israeli project to develop a ballistic missile and satellite launcher.[11]

In 1983, Saguy retired from the IDF and stepped down as head of military intelligence. He was replaced by Ehud Barak, a promising general who would go on to become Israel's chief of staff, defense minister and prime minister. Barak believed that Aman and Israel did not need satellites and could make do with aerial photos taken by reconnaissance aircraft, even if they were from within Israeli airspace and taken on an angle.

This was a potentially deadly blow to the program. Aman was supposed to be the satellite's main consumer, and if Barak was opposed to the development of a satellite, that meant there was no operational requirement. If there was no operational requirement, then there would be no satellite.[12]

The air force also wasn't overly excited by the prospect of an Israeli satellite. Air force commander Major General Avihu Ben-Nun recommended scrapping the program. He feared that the budget for the satellite would come at the expense of combat aircraft, which, he argued, were a higher priority for the country. Anyhow, Ben-Nun argued, Israel could purchase satellite imagery from the French or the Americans.

Another argument made by the air force was that it required real-time tactical intelligence. For satellites to be used in that way, they claimed, Israel would need about 20 of them in space at any given time, so they could switch off and continuously keep an eye on a specific area or operation. Such a network was obviously beyond Israel's budgetary capabilities.

But even without Saguy at his side, Eshed persisted. As he would later tell people: the word "impossible" was not in his lexicon.

As the launch date approached, the government faced a new

challenge. This was a covert program that was about to become public in a very noisy way. The air force base Israel planned to use for the launch—Palmachim—was located just south of Tel Aviv. The whole country and world were going to find out that Israel was up to something.

In addition, since the Convention on Registration of Objects Launched into Outer Space went into effect, in 1976, United Nations member states were supposed to update the organization with regard to satellite launches for its registry. Israel would have to comply even though until then it had officially refused to confirm reports that it was building a satellite.

Yitzhak Rabin, then Israel's defense minister, established a special committee to oversee the declassification of the satellite project. The committee's representatives were from different ministries, military units and IAI. Rabin decided that IAI's veteran and experienced spokesman, Doron Suslik, would do all of the media briefings and issue the press releases. The Defense Ministry would remain silent. The committee prepared a manual with a list of potential questions the media would be expected to ask. Suslik's answers played up the scientific achievement of the expected launch and downplayed the military dimension.

The military would remain in the background, with the hope that the launching of the satellite would be perceived as a scientific project without a military application.

* * *

On September 19, 1988, after a day's delay, Ofek-1 was launched into space, gaining Israel membership in the exclusive club of nations with independent satellite-launching capabilities: the US, Russia, France, Japan, China, India and the United Kingdom.

It was a historic day, and as planned, Israel played up the scientific significance of the launch.

Israel launches the Ofek Satellite from an air force base south of Tel Aviv in 2014. IAI

"This is a technical experiment . . . making Israel a partner in the top ranks of the modern technological era," Yitzhak Shamir, Israel's prime minister at the time, said a few days after the launch. "We should mainly consider the technological importance, and in that realm there is no doubt that Israel's international prestige has increased tremendously."[13]

Despite Shamir and Suslik's efforts, the global media focused on the launcher more than on the satellite. It was simple physics. If an Israeli missile had the energy needed to launch a satellite into space, it had the ballistic missiles needed to carry nuclear warheads across the entire Middle East.

And while the satellite did not have a camera—Israel even denied at the time that it was pursuing a spy satellite—Arab countries understood that it was just a matter of time before the Jewish State had the ability to keep an eye on their militaries all day, every day and everywhere.

The launch also sent a message to Washington. While Israel publicly voiced deep appreciation for the military assistance it received from the US, the satellite launch showed that there were limits to this dependence. As Begin had envisioned, an independent satellite launch meant an independent Israel.

Two years later, Israel successfully launched a second satellite, again without a camera. Now confident that its launcher worked, the Defense Ministry decided it was time to launch a real reconnaissance satellite.

Those who still needed some convincing with regard to the value of satellites got it in the form of 39 Scud missiles Iraqi president Saddam Hussein fired into Israel during the First Gulf War, in 1991. Without reconnaissance satellites, the IDF had no way of locating the Iraqi missile launchers or of providing early warning for Israeli civilians of incoming missiles. It had to rely on the US to give it a heads-up when a missile was launched.

At night, when the Scud missiles were fired into Israel, the IDF top brass would huddle in the Bor, the fortified underground command center below the Defense Ministry in Tel Aviv. These were tense moments. The IDF felt it needed to take action and drew up operational plans—including airlifting special forces by helicopter into the Iraqi desert—to locate and destroy the Scud missile launchers. Prime Minister Yitzhak Shamir shelved the plans and succumbed to US pressure on Israel to restrain itself. Washington feared that if Israel retaliated, the Arab coalition the US had pieced together against Iraq would collapse.

But one night, after more rockets had slammed into Tel Aviv, David Ivry, who at the time served as director general of the Defense Ministry, warned that this was just a taste of more to come.

"What we are currently seeing with forty-something Scuds is

nothing compared to what we will see in the future," he told the IDF generals present in the Bor.

Ivry knew a thing or two about the Iraqis. He was commander of the air force in 1981 and oversaw the bombing of the Osirak nuclear reactor. He was the one who convinced Prime Minister Begin that his pilots were capable of carrying out the strike.

Moshe Arens, the defense minister, himself a renowned aeronautical engineer, agreed with Ivry's assessment of the looming threat and immediately after the Gulf War summoned the IDF's research and development team to request an updated operational plan for reconnaissance satellites. "We need them now," Arens said.

But when the Defense Ministry launched a real reconnaissance satellite, in 1993, the launcher failed to reach space. The satellite was lost somewhere in the Mediterranean. In defense circles, Israel's satellites were being called "anti-submarine satellites."

It was up to Uzi Eilam, Eshed's boss as head of the Defense Ministry's R&D Directorate, to update the defense minister— Rabin had been reappointed to the post after elections—about the failure. When he had overseen the two previous successful launches, the minister's office was overflowing with VIPs and industry executives. This time, though, the room was empty.[14]

After the initial shock from the failed launch wore off, an independent committee of missile experts, who had not worked on the satellite program, was established to assess the failure. Five possible technical malfunctions were identified, mostly related to the launcher.

But Eilam and Eshed knew they were in bigger trouble. The launcher that went down was carrying the only available operational satellite. There wasn't a second one, let alone money to begin manufacturing a replacement.

They then remembered the "QM," an exact replica of the real satellite designed to serve as a test bed—a platform on which scientists could test other systems before installing them on the real satellite. The problem was that the QM was not built for space. It was not meant to be launched.

Two factions broke out within the defense establishment. The cautious faction was against launching the QM and instead recommended using a dummy satellite that would weigh 250 kilograms—the weight of the operational satellite—and would essentially be a test of the launcher, which had failed in the previous launch.

Eilam and Eshed were in the opposing faction. They argued for modifying the QM satellite so it could be launched. They wanted to improvise.

That was a risky position. Eilam and Eshed knew that another failure would not be tolerated and would bury the satellite project forever.

The temptation to play it safe, to wait and ensure that the new and modified launcher could succeed in reaching space, was very strong. Eilam and Eshed pushed back, however, and demanded that the launch be carried out immediately and with the QM. If not, they warned, Israel would lose the independent capability it had worked so hard to create.

They brought the risky proposal to Rabin, who, after some convincing, signed off.

After years of tests and modifications, launch day was finally set for April 5, 1995. When Eilam arrived at the launch site on the Palmachim Air Force Base, the launcher—decorated with a Star of David—was standing prominently, crowned by what was, until recently, a test satellite. There was no turning back.

Eilam took a seat behind the floor-to-ceiling glass inside the missile control room. On the other side, IDF officers were mak-

ing last-minute preparations and inspections of the various tele-
metric and radar systems deployed throughout the base to ensure
that the launcher stayed on course.

Off the coast, navy ships had finished clearing a corridor
in the Mediterranean to ensure that civilian ships would not cross
the path of the launcher in case it failed again.

"Five minutes and counting," a voice groaned over the intercom into the gallery. "Three minutes and counting."

At this point, there was only one man who could stop the
launch—the chief safety officer, an IDF colonel in the reserves
and the only person with the authority to abort the launch even
if told not to by the defense minister. Eilam checked; the colonel's hand was right next to the red abort switch. He was ready.

When the 10-second mark arrived, the missile went into
autopilot. The support beam at the launcher pulled back, and the
final countdown started. Suddenly, a voice blasted over the control intercom: "Stop! Stop!"

Eilam's heart sank, and with just a few seconds remaining, he
looked around the room, frantically trying to figure out what was
happening. The chief safety officer heard the calls, looked at his
various systems and removed his hand from the switch. And then
it happened: The missile's engine burst into flames and took off,
leaving behind a thick cloud of white smoke. The calls to stop had
been a false alarm.

Within minutes the first booster separated somewhere off the
coast of Libya. A few minutes later, another fuel tank fell into the
Mediterranean, not far from Algeria.

While the launch was successful, Eilam knew that they
weren't yet in the clear. He was waiting for the third stage,
when the satellite would be placed in space. Together with the
other VIPs, Eilam remained motionless on the other side of
the glass.

But after a few seconds, he finally heard the words he was waiting for: "We have proper separation. The satellite is in orbit."

The room erupted in cheers. Eshed had been watching the entire show at IAI's headquarters, where the satellite control room was located. While the satellite had been placed in space, it was now time to see if it was fully operational and could open its solar panels.

Eilam placed a call to IAI but could barely hear what Eshed was saying. "Haim," he shouted, "What's going on? Is it up?"[15]

It would take 12 more hours, but by the following morning, Ofek-3 had circled the earth eight times and had started taking photos. The resolution was better than anticipated. Planes, with their Israeli markings, could be made out clearly near Ben-Gurion Airport.

The success brought Israel immediate recognition, but more importantly, it quashed any opposition that might have still existed within the IDF. Everyone was now on board with Israel's satellite program.

✳ ✳ ✳

The launch of Ofek-1 on that day in 1988 was just the beginning. In the years since, Israel has grown into a satellite superpower. As with the other platforms it specializes in manufacturing, Israel has shied away from building big satellites and instead designs what are known as "mini satellites," which weigh about 300 kilograms in comparison with America's "mammoth" 25-ton satellites.

By 2014, with the launch of the Ofek-10, Israel had seven spy satellites in space, most of which use electro-optical sensors, cameras that can take high-resolution photos. The Ofek-9, launched in 2010, for example, carries the Israeli-made Jupiter multispectrum camera, which can discern objects as small as 50 centimeters from hundreds of miles away.

In addition to satellites with cameras, Israel also has two satellites that each carry a synthetic aperture sensor, or a radar system that can create high-resolution images. The advantage is tremendous. A camera cannot see through fog or clouds. Radars, however, work in all weather and can even see through camouflage nets.

With seven spy satellites in space, Israel now has the ability to operate them as a cluster. One satellite can bounce off the other and transmit images back to headquarters in real time, even when the satellite taking the photos is out of transmission range.

Israel's success in developing state-of-the-art satellites and associated payloads has caught the world's attention. In 2005, the French decided to capitalize on Israel's expertise and entered into a strategic partnership with IAI to develop a new satellite. Called Venus, the satellite was designed to study land resources, including vegetation, agriculture and water quality. In 2012, Italy ordered a reconnaissance satellite from IAI, paying $182 million. Singapore and India have also reportedly purchased Israeli satellites over the years.

Considering how Israel's satellite program began and evolved, this is an impressive accomplishment.

What was the secret to Israel's success?

What Eshed and Saguy displayed was not just a strong sense of innovation but also a persistent and stubborn flair—in other words, chutzpah. They had a lofty goal, which was to create an Israeli presence in space; they believed it was possible and refused to give up despite fierce opposition.

Eshed did what many visionaries in conservative organizations like militaries do: he bent the rules.

"I was surprised and still can't believe even today that Begin approved the plan," Eshed told us when we met at his Tel Aviv apartment overlooking the Mediterranean Sea. "There was a lot of opposition among the top military brass, and I just didn't think we would be able to convince the prime minister."

While breaking down hierarchal structures might seem at times to endanger an organization's ability to institute long-term strategic thinking, it can actually play a positive role by creating an atmosphere that tolerates a free exchange of ideas and criticism.

Eshed points to two key ingredients required for innovators to succeed in realizing their dreams. The first is the need to make sure that what they are proposing is humanly possible, or, as Eshed says, "does not go against basic physics."

The second ingredient is persistence but, more importantly, "thick skin, a strong spine and the ability to put up with insults and even the occasional rotten tomato thrown your way."

✳ ✳ ✳

Israel's satellites are controlled from a secret command center in the middle of the country by a unit known only by its number—9900. There, in a room with walls lined with plasma screens, soldiers track the satellites, send them on different missions and then wait patiently to receive their yield—the pictures.

Since its inception, 9900 has used its satellites to focus on strategic targets, enemies too distant for reconnaissance flights by spy planes or drones. Places like Iraq, Iran, Libya and elsewhere. In the first half of the 2000s, Israel's satellites were primarily focused on Iran, tracking developments of the ayatollah's nuclear program.

That all changed in the summer of 2006, during Israel's monthlong war against Hezbollah. Soldiers were sent into Lebanon without accurate intelligence. Their maps were outdated, and they were clueless as to where Hezbollah was hiding.

After the war, 9900 went through a shakeup and shifted its focus. The work paid off. In December 2012, the unit received an award for the intelligence it had gathered ahead of Pillar of Defense, the anti-Hamas operation that had ended just weeks earlier.

The change in focus was not simple. In the past, 9900 needed to follow Syrian military movements and keep an eye on Iraq. With enemies like Hamas in Gaza and Hezbollah in Lebanon embedded in civilian homes, the satellite operators had to work harder. Finding a rocket launcher buried in a schoolyard is far more complicated than tracking a Syrian armored division.

From their command center, 9900's troops build what the IDF calls "target banks" for its various fronts in Lebanon, Syria, Gaza and other places. "It reaches the point where I tell a pilot where exactly he needs to aim and shoot," a junior officer in the unit explained.[16] During Israel's recent operations in the Gaza Strip, satellite operators and analysts were assigned to the forward command centers of the IDF's various ground divisions. The idea was to create an intelligence pipeline that could work in real time between the unit, which gathers and analyzes the intelligence, and its consumers—the ground forces operating behind enemy lines.

Gathering the intelligence is only half the job. The other half is analyzing the imagery. For that, the IDF created a subunit of highly qualified soldiers who have remarkable visual and analytical capabilities. The common denominator among its members is just as remarkable: they all have autism.

The idea to recruit soldiers with autism came from Tamir Pardo, who until 2016 served as director of the Mossad, Israel's spy agency. He reached out to an Israeli NGO that specializes in integrating autistic youth into the workforce. "There has to be a

A model of Israel's Tecsar satellite that uses a radar instead of a camera to create high-quality images. IAI

way to utilize their capabilities for the benefit of Israel's intelligence community," Pardo said at the time.

The "special" soldiers were sent to a modified training course adapted for people with autism. In the beginning, the IDF was hesitant. While these are high-functioning autists, bringing them into a military unit still brought a risk. After a few months, though, the project's success exceeded even some of the more optimistic expectations. The soldiers specialize in identifying changes to terrain. If a bush moves a few feet or a building is slightly enlarged, they will pick up on it. To the average eye, these topographic changes might seem natural and be missed. But for 9900 they could mean that a rocket launcher or an arms cache is present but hidden.

This way of operating is unique. Most countries would auto-matically exempt autists from military service and would certainly not create a special training program for them. In Israel, though, this might have been expected. Autistic soldiers have unique capabilities, and Israel has limited resources.

✳ ✳ ✳

Satellites have revolutionized the modern battlefield. They have provided Israel with unprecedented intelligence-gathering capa-bilities, unmatched throughout the Middle East and most of the world.

But even as Israel has continued to bolster its presence in space, in 2009 it received a stark reminder that it is not alone in the region.

The Islamic Republic of Iran succeeded that year in launching Omid ("hope"), its first domestically manufactured satellite, into space. Like the first satellite Israel launched, in 1988, Omid did not carry a camera, and Iran claimed that its purpose was purely scientific. In 2015, Iran would launch a reconnaissance satellite.

Nevertheless, the 2009 launch was historic. Firstly, it dem-onstrated the progress the Iranian regime had made in the development of its ballistic missiles. If it could independently launch a satellite into space, it had the ability to launch a nu-clear warhead—once it built one—throughout the Middle East and even into some parts of Europe.

Iran was not the only Middle Eastern country pursuing a foothold in space. In 2007, Egypt's first spy satellite, jointly devel-oped with Russia, launched from there. But after three years, the satellite failed. Egypt continued with its ambitious space program and in 2014 launched a second spy satellite, also from Russia. After less than a year, though, the satellite mysteriously malfunc-tioned and lost contact with its ground station.

Iran's and Egypt's space activities mean that the satellite club Israel joined in 1988 is no longer as exclusive. The number of countries capable of independently launching satellites is growing, as are the threats to Israel's security. Israel is no longer the only country that can spy on its neighbors. It can be spied on now, too.

5

ROCKET SCIENCE

It was supposed to be like any other wedding—flowers, good music and a generous bar. The couple had invited about 300 guests, and the caterers had been setting up the outdoor wedding hall in Beersheba, the largest town in southern Israel, since early that afternoon.

At 4:00 p.m. that same day, a missile fired by the Israeli Air Force struck a silver Kia sedan driving down a residential street in Gaza City. The target was Hamas's elusive military commander Ahmed Jabari. It was November 14, 2012, and Pillar of Defense—the Israeli operation aimed at stopping rocket attacks from Gaza—had begun.

Jabari probably didn't even hear the missile as it streaked down at his car. An Israeli drone, hovering above, had been tracking him for a few hours, waiting for the right moment. As Jabari drove down the street, he passed a packed minibus. Once the car moved a safe distance away, the missile was launched, striking the car and killing its occupant. Debris flew everywhere.

After the assassination, the IDF Home Front Command, responsible for civil defense, issued a directive to close all schools within range of Hamas rockets. Under the guidelines, outdoor gatherings of 100 or more people were supposed to be canceled. The couple, however, decided to move ahead with their wedding plans. Yes, the ceremony, with the traditional Jewish chuppa, was supposed to be outdoors, but the hall was right nearby, and if a siren sounded, everyone could run inside. Despite the warnings of imminent retaliatory rocket attacks from Gaza, this couple was getting married.

Shay Malul, the wedding videographer, had met the bride and groom at 2:00 p.m. that day to begin filming. Two hours into the job, news about Jabari's assassination hit the airwaves. "I knew right away that there was going to be a *balagan*," Malul recalled, using the popular Hebrew slang for "chaotic mess." He called his wife and told her to pick up the kids and go straight home. He, however, planned to stay on the job. After all, he explained, filming had started. He couldn't just walk away.

By 7:30 p.m., most of the guests had arrived. There was a large buffet and a stacked bar. Everything was going as planned. But at 8:15 p.m. a warning siren sounded. A number of the guests started moving toward the wedding hall. Malul decided to play it safe and join them. But before going inside, he locked the camera on its tripod and randomly pointed it at the sky. That's when he saw the first of what looked like fireworks, bright lights streaking upward. When he saw another bunch, he ran inside the hall to take cover.

In a video Malul later posted on YouTube, 15 small bright dots are seen flying through the sky in different directions. They look like the beginning of a fireworks show. But, in fact, the fast-moving bright lights are Israeli Iron Dome missile interceptors launched at around a dozen Katyusha rockets fired seconds earlier from the Gaza Strip. In the video, the bright lights explode one after the other, intercepting the incoming Gaza rockets.

When retired IDF brigadier general Danny Gold, the man behind the development of Iron Dome, saw the YouTube video, the interceptions weren't what impressed him. There was something else that caught his eye. When the siren went off, some of the wedding guests fled indoors. Others remained outside and continued dancing to a cover version of "Sunday Morning," the hit song by Maroon 5, the popular American rock band.

This was the "Iron Dome wedding."

The development of Iron Dome is a mesmerizing tale that combines all of the characteristics Israelis are famous for—chutzpah, persistence, improvisation and plain old innovation.

Designed to intercept short-range rockets—which make up much of the arsenals of Hamas in Gaza and Hezbollah in Lebanon—Iron Dome has achieved stunning success rates. During the eight days of Pillar of Defense, in 2012, Iron Dome batteries shot down nearly 85 percent of the rockets heading toward Israeli cities. During Protective Edge, the anti-Hamas operation in the summer of 2014, Iron Dome achieved a 90 percent success rate.

This success is unparalleled anywhere in the world. No other country has a system like Iron Dome.

Israel's start in missile defense was by chance. In the mid-1980s, US president Ronald Reagan invited America's allies to join his Star Wars program, a missile defense system the US was developing to protect the country from Soviet intercontinental nuclear ballistic missiles. Yitzhak Rabin, Israel's defense minister at the time, recommended joining. Yes, it was true that Israel didn't really have anything to bring to the table, but Rabin's thinking was simple: Israel needed to strengthen its ties with the US, and being cooperative on missile defense could open new doors and opportunities. And cooperation required no immediate financial commitment.

To appear serious, though, Rabin ordered the Defense Ministry's R&D Directorate—known by its Hebrew acronym, Mafat, and basically the Israeli equivalent of America's Defense Advanced Research Projects Agency (DARPA)—to put some ideas on paper so that when the time came, Israel would have something to present the Americans. It wasn't a hard pitch to Israel's defense companies. They figured that if they showed something with promise, the Americans might throw massive funding their way. Even with a minimal investment, the Israelis could strike gold.

Within the IDF, Rabin's decision was met with skepticism. A panel of intelligence experts had recently evaluated the missile threat to Israel and determined that it was minimal and definitely not significant enough to warrant a massive investment in missile defense. Yes, Syria had an impressive arsenal of long-range Scud missiles, but that was basically the extent of the conventional threat against Israel. And while Syria's chemical weapons arsenal was considered a major threat, it could be minimized by distributing gas masks to the public.

There was also concern about Israel's joining a program that was clearly aimed at the Soviet Union. Moscow was the primary arms supplier of Israel's enemies. By joining Star Wars, Israel would be giving the Soviets an excuse to toughen its stance on Israel, supply additional advanced weapons to the Arabs and, at the same time, limit the number of permits it issued for Jews seeking to immigrate to Israel.

Nevertheless, Rabin ordered the R&D Directorate to move ahead. It in turn tapped Uzi Rubin to oversee the project. A talented young aerospace engineer, Rubin had proven himself on a number of classified defense projects and was known for being a no-nonsense manager.

Rubin dove right in, and within just a few months, Israel's defense companies came back with three solid proposals. One, crafted by a graduate of the elite Talpiot program, called for the development of a special chemical cannon that could fire 60-millimeter shells at unprecedented speeds. Another idea was to develop a missile defense test bed, or a laboratory where missile defense systems could be simulated on computers.

The last and most ambitious proposal was called Arrow. It called for the development of an interceptor that could shoot down incoming ballistic missiles just outside the atmosphere—a missile that could shoot down another missile.

It was a revolutionary idea. The developer, Dov Raviv, argued that it was critical to Israel's defense. Due to Israel's small size and lack of strategic depth, he said, all ballistic missiles deployed in the region could reach any target within the country. Israel, Raviv said, needed a system, based on a high-altitude interceptor, that could shoot down enemy missiles over neighboring countries and provide overall protection for Israel.

When the time came, Rubin led an Israeli defense delegation to Washington and pitched the three proposals. The Americans were surprised by the Arrow. But what really shocked them was Raviv's claim that development of the entire system would cost a mere $158 million. They thought it would cost a minimum of $500 million and would probably end up being much more. To Israel's surprise, the Pentagon decided to finance all of the projects.

The IDF top brass was not particularly happy with Washington's interest. The chief of staff, Lieutenant General Ehud Barak, sent a letter to the defense minister protesting the investment in Arrow. The argument was simple: the IDF needed tanks, fighter jets and navy attack ships. As the air force commander said during a General Staff meeting at the time: Missile defense doesn't win wars. Taking the offensive does.

At another meeting, Barak argued that the allocation of a budget for missile defense programs would undermine the country's chances of winning a future war. He urged Defense Minister Rabin to give all available money to the IDF to purchase offensive weapon systems. This way, we can finish the war quickly, Barak argued.

If Rabin insisted on obtaining a missile defense system, Barak continued, he could just buy the THAAD, a similar system under development in the US, whose price tag would be less than the development costs of the Arrow.

Rubin found an ally in David Ivry, the IAF commander who oversaw the bombing of Iraq's nuclear reactor and was now director general of the Defense Ministry. At one meeting with the IDF brass, opponents of the Arrow quoted David Ben-Gurion, who famously said that to survive, Israel needed to always take its wars into enemy territory. Investing in defenses for the home front, they claimed, was against the nation's ethos.

"That's true about Ben-Gurion," Ivry replied. "But you are forgetting that he also invested in building defenses for the Jewish community before the state was established. He knew that defense was just important as offense."

Toward the end of 1987, Israel received new intelligence that Syria was developing chemical warheads that could be installed atop the country's sizeable array of Scud missiles. This was a dramatic development. Israel had known about Syria's chemical arsenal, but until then, Syria would have needed to fly an aircraft into Israel to drop a chemical bomb, and the Israeli air force was confident that it would succeed in intercepting the Syrian plane. Now, Syria could simply lob a missile over the border.

This new intelligence came around the same time that Iraqi Scud missiles were slamming into Tehran during the Iran-Iraq War, leading to a mass evacuation of the city. Israel was able to see the devastating effects ballistic missiles had on civilian populations.

Nevertheless, nothing was moving in the IDF. Ivry decided to take action, and in March 1988 he wrote a classified letter to the defense minister, the chief of staff, the head of military intelligence and the commander of the air force.

"Surface-to-surface missiles are the greatest strategic threat to Israel and we need to take action," Ivry warned in the letter, personally accusing senior air force and military intelligence officers of neglecting the threat by underestimating its scope. "What they are saying is short of what the reality is."

The letter and its sharp criticism startled the defense establishment. The heads of the air force and military intelligence complained to Rabin. Ivry sent another letter a couple of days later, apologizing if his original letter was too personal. He refused, though, to back down from his demand that Israeli money be allocated for the Arrow.

While the IDF put up a fight, Rabin ultimately sided with Ivry and Rubin and approved a small but multiyear budget plan for the Arrow. "This is the program and there is nothing else," Rabin said at another meeting.

The program had its ups and downs, and in 1990, it seemed like it might be closed for good. But then, in 1991, the First Gulf War broke out, and Saddam Hussein fired 39 Scuds into Israel, paralyzing the country and forcing millions of Israelis into sealed rooms with gas masks. Israel was in complete panic. After the war, the Arrow program was pushed to the top of the nation's agenda with renewed vigor and, more importantly, increased budgets. The US also increased its financial commitment, even

An Arrow missile launcher on display during a joint American-Israeli missile defense exercise in Israel in 2016.   IDF

though Raviv's estimate was wrong. The program cost less than he originally predicted.

It would take another few years, but in 2000, the air force finally received its first operational Arrow missile battery, making Israel the first country in the world with an operational ballistic missile defense system. By then, Barak had also come around. Elected prime minister in 1999, he went on a tour one day of the IAI factory where the Arrow was manufactured. At one point he turned to Uzi Rubin, head of the Arrow project, and admitted his mistake: "You were right . . . I never thought we'd beat the Americans and be first in deploying a national missile defense system."[1]

For the first time, the country had a way to defend itself against Iraqi and Syrian missiles. Success, however, was short-lived. Israel didn't know it yet, but a new missile threat was brewing from an unexpected direction.

✳ ✳ ✳

Eli Moyal was sitting on the porch of his home in the southern
city of Sderot. Passover had just ended two days earlier, and
Moyal, Sderot's mayor, was enjoying the warm breeze sweeping
through his desert town. Suddenly, a massive explosion rocked
his windows. And then another. Moyal didn't think much of the
sounds until he saw smoke rising in the distance but still within
city limits. He jumped from his seat and dashed off toward the
smoke. When he arrived, he saw a hole in the ground and what
looked like a metal pipe sticking out.

"Don't tell anyone yet," a senior IDF officer who arrived at the
scene urged the mayor. "It seems like two rockets were fired from
Gaza and landed here, in Sderot."

It was April 2001, and Moyal couldn't believe what he had just
been told. "A rocket?" he asked. "In Sderot?"[2]

Hamas had named the rockets—Kassams—after the Kassam
Martyrs Brigades, the terror group's military wing, a notorious
and lethal organization behind countless suicide bombings and
shooting attacks perpetrated against Israelis. At the time, the
rocket's range was limited, and they barely flew a mile. By 2005,
Hamas had managed to increase Kassam's range to 10 miles. In
2006, it grew to 13 miles. In 2008, it jumped to 26 miles, and
in 2012 Hamas obtained Iranian rockets capable of striking Tel
Aviv some 40 miles away.[3] By 2014, more than 12,000 rockets had
been fired from the Gaza Strip into Israel, more than 1,000 of
which rained down on Sderot, making it the most visible symbol
of the rocket conflict and paralyzing a city that had been established to serve as a safe haven for Jews who fled Turkey and Iran
after Israel's establishment.

This was a surprising although not completely unexpected
threat. Since the 1990s, Hamas's trademark had been the suicide
bombings it carried out throughout Israel. There was the occasional drive-by shooting as well. But rockets were initially believed

to be beyond Hamas's competence. In hindsight, though, the use of rockets made sense for Hamas. Its role models—Syria and Hezbollah—had made a similar transition decades earlier, when they realized they could not compete with Israel's superior air and infantry forces. Missiles could bypass that superiority.

With the onset of the Second Intifada, in 2000, Israel tightened its hold on Palestinian parts of the Gaza Strip, controlling the sea and land crossings. Palestinians were literally locked inside. If Hamas wanted to attack Israel, it needed to come up with a new way. Rockets were the perfect solution.

One advantage of the first rockets fired into Israel in 2001 was in the availability of raw materials needed to manufacture them. They were lightweight and easily transportable and did not require complex launching systems. Any simple metal scaffold could suffice as a launcher. Pipes, sometimes taken from street-

The IDF Home Front Command holds a drill simulating missile attacks on central Israel.   IDF

lights, could serve as the rockets. And, most importantly, Hamas didn't need to cross an Israeli checkpoint or bypass an IDF patrol to attack. The rockets just flew over them.

By 2005, Hamas was receiving its missiles from two primary sources. The short-range Kassams and Katyushas were designed and manufactured locally in the Gaza Strip. Longer-range missiles were smuggled into Gaza through a network of tunnels the terror group had dug and operated along the border with Egypt, a small nine-mile strip of land called the Philadelphi Corridor. Sometimes the missiles were too big to fit into a tunnel in one piece, so they were disassembled before being smuggled into Gaza. There, Hamas engineers reassembled them.

What Moyal saw that day was not Israel's first taste of rocket fire. Northern Israel had been shelled before by Hezbollah from bases in Lebanon. But those two Kassam rockets carried a startling message: the rocket threat to Israel was spreading, and the country was once again, defenseless.

It took Israel time to fully understand the extent of this new threat. In 2001, Israel still had settlements inside the Gaza Strip, and while they frequently came under attack, it was mostly from mortar shells or the occasional lone gunman who infiltrated a settlement. It also took time for the rocket fire to intensify. In 2001, only four rockets were fired into Israel. By 2002, it was 34. And in 2003 the number was 155. The trend was becoming clear.[4]

Israel thought it already had something of a solution. In 1996, then–Prime Minister Shimon Peres signed an agreement with President Bill Clinton calling for the joint development of a missile defense system based on a laser called Nautilus. The perceived threat then was the Katyusha rocket fire from Lebanon, but the Israeli government figured the system could easily be applied to other areas as needed. The problem was that the development of the laser was taking longer than expected, and it was unclear if it would ever really work.

The big breakthrough came in 2004, with the appointment of Brigadier General Danny Gold as head of the Defense Ministry's Research and Development Department. Gold was drafted into the air force as a radio engineer but had a keen knack for developing weapons. In the 1990s, when he was a colonel in charge of weapons development in the air force, he took a sabbatical and enrolled in two doctoral programs at Tel Aviv University, one in business management and the other in electrical engineering. He completed both within two years.

Shortly after taking up his new post, Gold decided that the emerging rocket threat from Gaza would be one of the R&D Department's main focuses. His decision was based on intuition. Yes, the threat from Gaza was just emerging, but Gold believed it had the potential to turn into a challenge of national and strategic proportions.

Gold first went the traditional route and submitted a budget request for a technological review through the standard bureaucratic channels. Everywhere he went, though, he heard the same answer: "Forget about it. There isn't money." When pressed, the generals he met used one of, or a combination of, four excuses: your idea will ultimately fail; it will take 20 years to develop a solution; it will cost billions of dollars; and by the time a rocket defense system is operational, it will no longer be relevant. Israel, these generals argued, needed to invest in offensive capabilities, not in a stronger defense. It was the Arrow saga all over again.

Gold persisted. Such a missile defense system, he argued, would actually allow Israel to be more aggressive on the battlefield. If Israel's civilians were protected, there wouldn't be pressure to end a conflict due to rocket attacks. He also warned of the potential economic ramifications of a large-scale missile attack on

Israel. "If we succeed, the system will not only protect people but it will also give the government time to think before needing to respond in the event of an attack," he argued.

Despite overwhelming opposition, Gold decided to move ahead. While his own department's budget was small, he managed to shift things around a bit and set aside a tiny budget for the first step, the establishment of a development team.

Before the team got to work, Gold went to military intelligence to hear its predictions regarding the evolving rocket threat against Israel. The intelligence analysts told him it would take years before Hamas had rockets sophisticated enough to pose a strategic threat to Israel's home front. As a result, they said, there was no reason to rush into developing a system.

"What difference does it make?" Gold asked the analysts. "Hamas will eventually get there, even if it takes a few years, and anyhow it will take us time to develop the system. Now is the time to start." Gold knew about the Nautilus laser project and the hundreds of millions that had been poured into its development. But he had also reached the conclusion that it wasn't working and probably wasn't going to.

So in August 2004, Gold issued a request for information (RFI) to Israeli defense companies, asking them to present new ideas for a rocket defense system. Within weeks, his team received no fewer than 24 proposals, ranging from kinetic interceptors like the Arrow to variations of the Nautilus laser to high-speed rapid-fire cannons. The top military brass was extremely skeptical that it was even possible to shoot down rockets fired from the Gaza Strip, especially when they were flying toward places like Sderot, only seconds away.

But Gold's team evaluated all of the proposals.

One system was based on the Vulcan Phalanx, a high-speed cannon designed by General Dynamics to protect naval ships from incoming anti-ship missiles. The Americans were adapting the system for land use to protect forward-operating bases in Iraq from rocket and mortar fire. Israel's problem with using the Vulcan Phalanx was simple: the cannon fired around 4,000 rounds a minute at incoming rockets. The rockets were fired from Gaza, which meant the rounds would be fired toward Gaza. How could Israel justify firing that number of rounds at the Gaza Strip because of a couple of mortars and rockets?

Then there was the upgraded version of the Nautilus, now called Skyguard. Gold's team evaluated the system but found it unsuitable as well. There were three reasons: the laser could not work in cloudy weather, the system was too big to move around as quickly as needed and it could not effectively intercept rocket barrages. It was also, apparently, years away from becoming operational.

The team traveled to the US, France and Germany to see some of the systems in action, but nothing seemed to be close to what they were looking for. Gold had laid out clear guidelines for what he wanted, but one factor in particular—it needed to be cheap—was critical. It would take a few more months, but by mid-2005, Gold and his team were convinced that they had found the right system. Rafael, a government-owned company, world renowned for its air-to-air missiles, had put together a concept based on a new rocket interceptor.

The idea was quite innovative. Called Iron Dome, the system consisted of three main components. The first was an interceptor, a missile that could intercept incoming enemy rockets. The second component was a powerful radar to detect the launching of rockets from enemy territory. The third component was a battle management system, based on advanced algorithms, which would be able to predict the rocket's trajectory and determine where it was going to land within mere seconds of its being launched. This way, the IDF would be able to warn residents of a specific target area but also hold back from wasting interceptors on rockets landing in open fields. Only rockets projected to land in populated areas would be targeted.

In addition, since there would be only a few seconds between launch and interception, the system had to know how to work automatically and without human intervention. Lastly, and possibly most importantly, each individual interceptor had to be cheap. "If the interceptor costs a million dollars, even if it works, the army won't buy it," Gold said. And also, he pointed out: "If it's not cheap, the enemy will simply bankrupt us with rocket barrages."

To move the process along, Gold did something beyond the usual Israeli chutzpah. He decided to break the rules. Violating numerous regulations, Gold gave Rafael a green light in August 2005 to begin developing the system. Gold then went a step further, doing something that only the IDF chief of staff or minister of defense was authorized to do: he ordered Rafael to begin full-scale production the moment it was ready. He also set a schedule for the system's eventual delivery. "We need to get to an operational capability as soon as possible," Gold told his team.

This was a risky move. Usually, when developing a new weapons system, the process goes like this: the IDF sets the criteria for the new weapon and then R&D people like Gold work on developing the concept. Then the R&D Department issues a tender, and defense companies are given time to submit proposals. Here, Gold was skipping over the IDF rulebook. His actions didn't go unnoticed. In 2009, Israel's state comptroller issued a scathing report on the Iron Dome project, slamming Gold for violating military regulations. Gold, the comptroller concluded, "took upon himself authority reserved for the chief of staff, the defense minister and the cabinet before the project was approved by the relevant authorities." It didn't make a difference, though. By the time the report came out, Iron Dome was already a success.

In a meeting in 2005 with Rafael's chairman of the board, Ilan Biran, Gold confessed his biggest problem. "There is no government funding for this program," he said. "But I have $5–$6 million of my own research budget that I can commit to the project if you match it."

Biran said he was willing to explore the possibility, and Gold tried to reassure him. Whatever happened down the road, he told Biran, he would obtain the necessary budget for full-scale development and production. Then, to make sure he would be able to stand by his word, Gold did something highly unusual for a military officer: he contacted a private Israeli venture capitalist based in the US and asked him to prepare a $50 million investment. Gold knew the businessman from a defense start-up the two— Gold at the time represented the air force—had invested in a few years earlier. "I can't tell you what I need the money for but be prepared that I might call and ask for it," he told the investor.

Biran asked for a few days to consult with the company's engineers and missile experts. He convened a meeting of his top missile developers to get an answer to a simple question—can it be done?

All eyes were on Yossi Drucker, a veteran missile developer who had worked at Rafael since the late 1970s and was head of the company's missile department. Drucker and his team already had seven different missile projects under their belt. They were Rafael's missile team.

Rafael developed its first air-to-air missiles, or AAM, in the 1950s, but it wasn't until the 1973 Yom Kippur War that it started to see success. During the war, the Shafrir AAM shot down close to 100 enemy aircraft. Five years later, Rafael made a huge technological leap with the introduction of the Python 3.

The difference was that the Shafrir had to be lined up directly behind an enemy aircraft to hit its target. With the Python 3, the IAF could shoot down enemy planes from different angles and positions. During the First Lebanon War, the Python 3 shot down nearly 40 enemy aircraft. The missile continued to improve, and by 2006, the IAF was mostly using the Python 5, an AAM that could lock onto targets after it had already been launched, meaning that the pilot could shoot down enemy planes without even seeing them.

"The idea is this," Drucker explained to the team one day. "If we can use a missile to take out an aircraft, then we should be able to get our missile to intercept another missile."

Not everyone thought it would be that simple. When it comes to aiming at a plane, the missile has a long and wide target to lock onto. But when the goal is to shoot down a 170-millimeter rocket, the target the interceptor needs to lock onto is tiny. Blowing something up within a few feet was not going to be enough. The missile needed to get really close. Shooting missiles into the sky during a war was complicated for another reason as well. The air force needed the skies to be clear so its planes could take off and land. Now, it would need to worry about Israeli interceptors flying in various directions. Israel's skies were about to get very crowded. "It won't be easy, but we can do it," Drucker said. Biran gave Drucker the green light to assemble a team of engineers and scientists and to start work.

Due to the shortage in funds, Rafael and Gold's team needed to cut costs and get the necessary materials and components as cheaply as possible. One question that came up, for example, had to do with the loading of missile canisters—each came with eight missiles—on the launcher. A member of Gold's team was on his way to work one day and watched how a garbage truck used a forklift to pick up large street dumpsters. He contacted the company, and a few weeks later, a similar forklift system was delivered to Rafael headquarters.

Work continued, but without full government support, the deadline kept getting pushed back. Then, in the summer of 2006, everything changed. On July 12, 2006, Hezbollah guerrillas crossed into Israel and attacked an IDF border patrol. Two reservists were abducted. In an attempt to cut off the infiltrators' escape, a nearby Merkava tank—the pride of Israel's domestic defense industry—stormed across the border. It rolled over a massive bomb and was blown to smithereens.

The combination of the kidnapping and the loss of four soldiers in the tank shocked the country. Prime Minister Ehud Olmert decided to retaliate, sending the country into war, its first in almost a quarter-century.

The war would ultimately provide Israel with a decade of quiet along its northern border, but it also opened Israel's eyes to the true dimension of the rocket threat it faced. In just 34 days, Hezbollah fired a staggering 4,300 rockets into Israel, an average of more than 120 a day. The Israeli public was traumatized. Tens of thousands of people fled their homes. Northern Israel became a ghost town.

A few days after the war ended, Defense Minister Amir Peretz convened a meeting in his Tel Aviv office to review missile defense options. The war had been traumatic for Peretz, and his political career was hanging in the balance. A veteran union leader, declared social reformer and head of the socialist-leaning Labor Party, Peretz had asked to serve as finance minister after elections were held a few months earlier, but Olmert was concerned that the appointment would shake up the economy. Olmert's ad-

A Hezbollah missile launcher photographed before being bombed by the Israeli Air Force during the Second Lebanon War in 2006.   IDF

visors warned against appointing Peretz to the Defense Ministry, but Olmert had made up his mind. Anyhow, he told them, as prime minister he would be able to supervise Peretz from above.

With zero defense experience beyond his mandatory military service, and coming in after years of former IDF generals as defense ministers, Peretz was viewed suspiciously within the IDF and beyond. But he knew a thing or two about rockets. He was a longtime resident and former mayor of Sderot; his family had been the target of rocket attacks from the Gaza Strip for nearly six years. At last, there was an opportunity to do something about it.

"Iron Dome is the most important project right now," Peretz said at the meeting. "We should consider speeding it up despite the costs."

Not everyone agreed. The deputy chief of staff, Major General Moshe Kaplinsky, who until the war had been the front-runner to become the next chief of staff, urged patience. "It's easy to make decisions today, after everyone spent a month in the bomb shelters," the veteran IDF general said. "These decisions, though, can take us places we don't want to go." Peretz dismissed the warning and summed up the meeting with an order to Gold to speed up development of a rocket defense system.

In the following weeks, Olmert was also briefed on the system for the first time. The discussion didn't go as well as Gold had anticipated. Nearly the entire IDF top brass opposed the project, and Olmert, under pressure, refused to divert government funds.

The war's impact was also felt back at Rafael. The company's missile factory is located in northern Israel. Many of the engineers and workers live in nearby towns and had to either flee south during the war or spend the 34 days holed up in bomb shelters. Even without government funding, Iron Dome suddenly found its way to the top of the company's agenda.

Drucker, head of the missile department, knew who he wanted as the project manager, but the problem was that the man—Uzi—had just left for a trek through Chile, a vacation he had planned for months. After a day or two, Drucker managed to reach Uzi by phone. "Come home," he said. "We need you." Uzi asked for a couple of days to consider the offer and mostly to convince his wife that they would need to cut their dream vacation short. She agreed, and within a week Uzi returned to Rafael.

It took a few more days to get Uzi on board and behind the concept. But he quickly told his staff: "The word 'impossible' does not exist when it comes to this project." Gold, meanwhile, was still trying to maneuver amidst the opposition inside the IDF. In November 2006, he again bent the rules and unilaterally contracted Rafael to continue full-scale production. Gold was skipping over key procedures, as the IDF had not yet completed an internal review aimed at determining which unit would operate the system and what exactly it would need to do.

In early 2007, Peretz came to Rafael's missile factory to meet the engineers and see the assembly line. He was about to throw his full ministerial weight behind Iron Dome and wanted to see where the millions of dollars he was considering allocating would be going.

The Rafael missile factory is one of Israel's most secure facilities, hidden behind electronic fences and armed guards among the rolling picturesque Galilee hills in northern Israel. Some of the military's most sensitive missiles and bombs are invented and manufactured here. At the entrance to the main administration offices, there is a large hall with models of some of the missiles Rafael has developed over the years, an impressive testament to the company's technological capabilities.

At the time of his visit, Peretz's popularity ratings were at their lowest. Reservists, who had returned from Lebanon frustrated with the government's handling of the war, had set up a massive protest camp in Jerusalem and were calling for the defense minister's resignation and the opening of an official state inquiry into the war.

If that wasn't enough, a few weeks earlier, Peretz was at a military exercise in the Golan Heights. The photograph that led the media the next day showed him looking through binoculars with the lens caps still on. He became an international laughing stock.

What makes Rafael unique is that unlike high-tech companies, it considers age a value, not a liability. Walking through the factory's halls, Peretz saw engineers in their seventies sitting next to recent graduates of the Technion. The older engineers were working with pencils and yellow legal pads. The younger ones were tapping away on laptops.

Peretz was shown the different missiles, received a short explanation about each of them and then went to the assembly line, where Drucker showed him around.

"I expect you to work three shifts a day," Peretz said.

"No," Drucker said to the stunned defense minister. "We are already working one shift. It's 24 hours long."

What Peretz didn't know was that the company was keeping its doors open on Saturday, Shabbat, the Jewish day of rest. It had received special permission from rabbis to keep the assembly line going. Lives were in danger, and Iron Dome was needed to save them.

But while Gold, Peretz and Rafael were moving forward with Iron Dome, there were still vocal critics in the defense establishment. One of the major opponents was a former senior executive at Rafael who continued to lobby for Skyguard, the upgraded version of the laser system formerly known as Nautilus. The campaign was harsh. Articles appeared almost daily against Iron Dome, claiming it wouldn't work against missile barrages and that even if it did, it would bankrupt the country with a price tag of between $50,000 and $100,000 per interceptor.

In June 2007, Peretz was replaced as defense minister by Ehud Barak, the former prime minister and IDF chief of staff. A few weeks after taking office, Barak asked Gold and his R&D team to once again review Skyguard and see if there was a way to develop the laser system parallel with Iron Dome. Gold opposed the new review but had no choice but to cooperate. Rafael executives were nervous that the review would end in favor of Skyguard and that their work and money would go down the drain. "Don't worry," Gold told them. "Our system is the only viable one."

But even as Gold and Rafael plowed forward, the government

had to confront a new challenge: even if the system was successful, how would Israel afford enough batteries and interceptors to protect its borders?

The answer, everyone agreed, rested 6,000 miles away—in Washington, DC. The question was how best to approach the Americans. A meeting at the Defense Ministry ended with a decision to submit an initial request for technical cooperation. It landed on the desk of Mary Beth Long, assistant security of defense for international cooperation under Defense Secretary Robert Gates.

To assess the request, the Pentagon sent a team of experts to Israel to meet the developers. They returned unimpressed. For starters, the US team believed Israel was underestimating Iron Dome's true cost and that a protracted conflict with a major missile onslaught on Israel would, in fact, bankrupt the nation. The American engineers also thought Israel's prediction of a high interception rate was way off and that in reality Iron Dome would, in the best case, be able to take out about 15 percent of rockets fired its way.

"This cannot be done," the team members told Long.

A few weeks later, a Defense Ministry delegation came to the Pentagon to meet Long and her team. It was led by Amos Gilad, director of political-military affairs at the ministry and a former veteran IDF intelligence officer.

Long got straight to the point. "Why are you here asking for this now?" she challenged the Israeli team. "We just completed a package in which you got more money than ever for military aspects." Long was referring to the signing, just a few months earlier, of a new memorandum of understanding between Israel and the US, setting US defense aid to Israel at $3 billion a year for the next 10 years. It was the largest foreign military aid package ever.

The new aid was the result of years of negotiations between Israel and the US, but the war in Lebanon had played a key role.

The threats to Israel were increasing, and the George W. Bush administration knew that if it wanted Israel to take risks toward peace with the Palestinians, it needed to provide the country with a sense of security.

Long figured that any money Israel wanted to develop Iron Dome should come out of the $3 billion it was going to receive each year. The problem was that the Defense Ministry already had other plans for the US money—mostly to buy combat aircraft and replenish missile stockpiles, exhausted during the war. Long didn't like the answer and urged Gilad and the delegation to rethink their request. "Before you ask my department to make hard budgetary decisions, I want to at least see that you've made hard budgetary decisions," she said. "Don't come back and tell me I need to go find this money in my system when I don't see any evidence that you've tried to do that with yours."

There was another problem that posed something of a bureaucratic hurdle. In the Pentagon, Long's office could fund only proven weapons programs, not systems still under development. That meant Iron Dome would have to be evaluated by another Pentagon office. For Israel, that meant more time lost. Long could have rejected the proposal outright, but she didn't. She decided to give it a chance, albeit a small one. Instead of killing the idea, she asked Brigadier General Robin Rand, head of Middle East policy at the Pentagon, to establish the Short-Range Rocket Defense Working Group with the Israeli Defense Ministry to iron out the differences. At Rafael's missile center, in the North, engineers loaded the Iron Dome launcher onto a truck for the long drive down to the company's missile range in the South, near Israel's border with Egypt. It was time for a "flyout," a flight test of the interceptor called Tamir, to make sure that it could maneuver as designed. Until then, all of the tests had been computer simulations. This would be the real thing.

The operator began the final countdown: "5, 4, 3, 2, 1." All eyes—of the Rafael development team, IDF officers and Defense Ministry officials—were glued to the bank of screens in the room. One screen showed a color video feed of the launcher. Another showed a fuzzy infrared image so the developers could track the interceptor even as it passed through clouds.

But when the operator pressed the launch button, nothing happened. He pressed again, a little harder, but still—nothing.

Drucker and his team feared the worst: that their system was a total flop. With the media closely following the development of Iron Dome, this kind of news could bury the project before a single missile got off the ground.

Drucker didn't have much of a choice. He loaded the launcher back on the truck and returned it to the Rafael plant. A few days later, the engineers identified the problem. It turned out that a cable had accidentally dislodged and caused the system to malfunction. Two weeks later, they were back at the range. The operator pressed the launch button, and the interceptor, Tamir, took to the sky.

Despite the successful test, Iron Dome was not yet in the clear. In 2009, the night before the first live interception test, the engineers discovered a bug in the system's software.

"We should delay," some of the team members told Gold. "All of the top brass will be there. We will look bad if we fail." Gold thought for a few minutes and decided to go ahead with the test. "Even if it fails to intercept, we need to learn," he told his people.

The next morning, the group again drove to the test range. A little before 11:00 a.m., the mock Katyusha rocket took off. Everyone watched the Iron Dome operator's screen and saw the radar immediately detect the launch. Within seconds a Tamir interceptor was launched. Everyone held their breath until an explosion rocked the building. Iron Dome had worked. It had shot down its first Katyusha. The ensuing applause and jumping up and down almost brought down the building.

The US team, led by General Rand, was following these developments with great interest. Subsequent tests proved that the system's interception rate was significantly higher than initial US estimates. Iron Dome could shoot down at least 80 percent of incoming rockets. Nevertheless, the Pentagon was reluctant to throw money into the project. That would need to wait.

In July 2008, Illinois freshman senator Barack Obama arrived in Israel. This was Obama's second visit to the country but his first as a presidential candidate. He arrived for two days as part of a whirlwind tour that also took him to Kuwait, Jordan, Germany and France. Compared with his opponent, veteran senator John McCain, Obama had barely any foreign policy experience. The trip was supposed to give him some desperately needed credibility.

Obama made the mandatory stops at the Yad Vashem Holocaust Museum and the Western Wall. But he also paid a visit to Sderot, the southern city most associated with Hamas rocket fire. He went to the local police station and stood in what is known as the "rocket morgue"—a large yard full of remains of rockets that have struck the city. That spot was the background for a prepared statement about the need to stop Iran's nuclear program.

Afterward, Obama was asked by a reporter whether he would accept a situation in which a city in the US was under constant rocket fire, as Sderot was.

"I don't think any country would find it acceptable to have missiles raining down on the heads of their citizens," Obama said. "If somebody was sending rockets into my house where my two daughters sleep at night, I'm going to do everything in my power to stop that. And I would expect Israelis to do the same thing."

The visit to Sderot especially moved the future president. Afterward, Obama told his aides that should he win the election, his administration would need to help Israel find a way to boost its defenses against rocket fire from the Gaza Strip.

But with the US elections months away, Israel's request for funding was gathering dust at the Pentagon. No one expected a new funding initiative when the sitting president had just a few months left in office. Even after the elections, Israel knew there was no point in immediately approaching the Obama administration. It needed time to settle in.

April 2009 was the turning point. Georgetown professor Colin Kahl, a foreign policy expert, was appointed deputy assistant secretary of defense for the Middle East. His task was to oversee US military policy in the volatile region and help come up with ways to promote stability. With the peace process in deadlock, Obama was determined to get Israel and the Palestinians back to the negotiating table. Pressure was mounting on Prime Minister Netanyahu to agree to a freeze on settlement construction. The US needed leverage to make all of this happen.

In the meantime, Defense Minister Ehud Barak came to Washington and gave the Pentagon a paper outlining the basic security requirements Israel would need in place before it could even contemplate a withdrawal from the West Bank and the establishment of a Palestinian state. The bottom line was an Israeli concern that a withdrawal from the West Bank would lead to rocket attacks on the center of the country, as had happened after Israel withdrew from the Gaza Strip a few years earlier. It was then that Kahl found the Iron Dome request on his desk and experienced what he later described as the "lightbulb moment."

Kahl took the idea to Dan Shapiro, head of Middle East policy in the National Security Council at the time and soon to be appointed the US ambassador to Israel. "Iron Dome shows promise," Kahl told Shapiro. "If it works, I think it would facilitate the Israelis taking greater risks to sign on to a two-state solution."

Kahl and Shapiro agreed to send a new team of missile defense experts to Israel to review the system. It was a controversial move. At the time, the US was trying to interest the Israelis in the Vulcan Phalanx, the high-powered cannon that had already been ruled out by Gold and his R&D team. But Kahl sent the new team, which came back singing the Iron Dome's praises.

In June 2009, Kahl traveled to Israel for his first official visit. The IDF flew him by helicopter to the border with Lebanon, where he was briefed on Hezbollah and the organization's military buildup since the 2006 war. He then flew south, to the border with Gaza, where he received a briefing on Hamas's growing rocket capabilities.

He was struck by Israel's lack of strategic depth and how close towns and cities were to the threats brewing in the North and South. When Kahl returned to Washington, he drafted a memo recommending that the White House immediately authorize $200 million in Iron Dome funding. His argument was simple: Israel wanted security assurances, and the Iron Dome could provide them. The president would get to kick-start the peace negotiations, and Israel would get an additional layer of security.

Iron Dome went operational in March 2011, when the IDF deployed its first battery outside Beersheba. It didn't take long for the system to see action. On April 7, Iron Dome shot down its first rocket, and within a few days it intercepted eight more.

Beyond saving Israeli lives, Iron Dome proved to be a game changer. In recent IDF operations in Gaza, it has intercepted about 90 percent of the rockets fired its way. In 2012, the IDF did not send ground forces into Gaza, and in 2014, it sent troops in for only an isolated small-scale operation against tunnels. With the Iron Dome shooting down most of the rockets flying at Israeli cities, the government had "diplomatic maneuverability"—the ability to think before responding, a precious commodity in times of crisis.

Systems the scale of Iron Dome take, on average, seven years to design and manufacture. Iron Dome took just three. How did Israel develop a revolutionary system like Iron Dome in such a short time?

Part of the answer is that Israeli military officers and businessmen like Gold tend to be less risk-averse than their counterparts in other Western countries. Naftali Bennett, Israel's minister of education, told us about one of his experiences as a high-tech entrepreneur. Bennett had served as an officer in two elite IDF units—the General Staff Reconnaissance Unit (known as Sayeret Matkal) and Maglan, both of which specialize in covert operations deep behind enemy lines. At the age of 21, Bennett was already leading 100 soldiers on operations in Lebanon.

A few years after his discharge, Bennett found himself at the entrance to a bank in New York, where he and his fellow high-tech partners were about to make their first business pitch. They had come with a new anti-fraud software that, a few years later, would be sold for $145 million. "Everyone was nervous," Bennett recalled, "and I said: what's the worst that could happen? They'll say no, right? But at least no one will die. No one will step on a mine." To some extent, this was also Gold's approach with Iron Dome. The worst that could happen was that Gold would fail and pay with his career. He felt that the potential reward was worth the personal risk.

But that still doesn't explain why Gold violated military regulations—why he didn't wait and take the safer route. Israel, Gold told us when we met one day in Tel Aviv, didn't have the luxury of waiting. It needed to survive. "We knew there were thousands of rockets in Gaza and tens of thousands more in Lebanon," he said. "What were we going to do? Wait some more?"

Israel's development of missile defense systems has changed modern warfare. Israel is the only country in the world that has used missile defense systems in times of war.

For Israel, systems like Iron Dome and Arrow are about more than saving lives. They give the country's leadership the ability to think before retaliating against rocket attacks. They provide the IDF with the ability to protect its bases and ensure operational continuity, to keep planes taking off and landing even if missiles are being lobbed at runways.

The IDF is working on a third system, called David's Sling, which will be used to intercept rockets that are too big for Iron Dome but not big enough to be intercepted by the Arrow.

Additional countries around the world have also invested in missile defense systems—the US, Japan and South Korea, among others. None, though, have created a multitier architecture with different systems, as Israel has.

But even as Israel has developed and deployed these systems, its ultimate prayer has not been answered. When Iron Dome was first deployed, there were some defense officials who predicted that if it worked, Hamas would abandon its rockets. It would realize that rockets were no longer effective and would stop spending money on amassing larger arsenals.

This has not happened. Israel's enemies are still amassing rockets and missiles at a dizzying pace. According to recent Israeli intelligence assessments, Hezbollah's arsenal is the most impressive. In the span of 10 years, mostly due to assistance from Syria and Iran, it has succeeded in increasing its numbers from 15,000 to more than 100,000 rockets and missiles that are capable of striking anywhere inside the State of Israel. Hamas is believed to have about 10,000.

The threat to Israel is not just from the quantity of missiles but also from their improved quality. The IDF's Military Intelligence Directorate calls this transformation "Fire-by-6," a reference to the six changes Hezbollah's arsenal has undergone in recent years.

Today, Hezbollah has more missiles with longer ranges, larger warheads, greater accuracy, and the ability to launch from deeper inside enemy territory—not just along the border—and in some cases even from within fortified and underground silos.

For example, the M-600 made in Syria has a range of 200 miles, carries 500 kilograms of explosives in its warhead and is equipped with a sophisticated navigation system, giving Hezbollah an unprecedented level of accuracy. Israel believes Hezbollah has hundreds of M-600s stored in underground silos and homes throughout central and southern Lebanon.

This constant and growing danger is what helps foster innovation in Israel. It makes people think in order to survive. "We can either innovate or disappear," explained Arieh Herzog, a former head of Israel's missile defense agency. Herzog was born in 1941, two years after the Nazi invasion of Poland. Following his father's murder by Nazis, Herzog's mother disguised herself as a Christian farmer and fled with her son to Hungary, where they hid out for the duration of the war.

When US Missile Defense Agency officials would visit Israel, Herzog, as their host, would first take them on a tour of the Yad Vashem Holocaust Museum before getting down to business. "After seeing what can happen to our people, you understand the importance of ensuring it doesn't happen again," he explained. "This isn't some virtual threat. It's our daily reality."

6

INTELLIGENT MACHINES

Few were privy to the secret. After years of hunting, the Shabak—the super-secret Israeli internal security agency—succeeded in locating the most wanted and elusive man in Gaza: Mohammed Deif. It was one of those moments that rarely occur in an intelligence officer's career. The neon lights came on in the special command center at Shabak headquarters on the outskirts of Tel Aviv. Officials took up their positions around the oval desk in front of the big plasma screens. Intelligence from informants, drones and satellites started flowing in.

At air force headquarters, the dimensions of the building in which Deif was reportedly hiding were being analyzed, and experts were carefully selecting the bombs that would need to be dropped. They had to be small enough to limit collateral damage but large enough to get the job done and kill the man who for years had escaped death. It was a race against the clock. Deif never stayed long in one place. Nevertheless, the intelligence had to be checked and checked again. After what seemed like an eternity, the green light was finally given, and a pair of air force fighter jets took off toward their target—a small apartment building in the Gaza City neighborhood of Sheikh Radwan. It was August 19, 2014, and Israel was in its fifth week of fighting with Hamas in the operation known as Protective Edge.

Deif was not just any wanted man. He was at the top of the Hamas pyramid and had evaded capture for nearly 20 years. This was not Israel's first attempt to kill him. The last time had been in 2006, but Deif had survived, albeit badly injured. Somehow, he always managed to slip away.

After weeks of fighting, a rising casualty count and incessant rocket attacks across the country, Deif's elimination was supposed to provide Israel with a desperately needed boost of morale. Getting accurate intelligence on Deif's whereabouts was in itself a big deal. No one except for a small, highly compartmentalized group of bodyguards was supposed to know their commander's exact location. The fact that Israel had located him was itself a major coup.

It didn't take long after the bombs struck for rumors of Deif's assassination to spread like wildfire. One of Deif's wives, together with his eight-month-old son, were identified among the dead. Another body was found, but no one could confirm that it was Deif's. Either way, it was as if an arrow had struck at the heart of Izz ad-din al-Kassam—Hamas's military wing. Deif had been its supreme operational authority, a spiritual figure and one of the most prominent symbols of the Palestinians' decades-long struggle against Israel. If Deif was dead, he would leave a large power vacuum.

Assuming their commander was dead, Mohammed Abu Shamala, head of Hamas's Southern Division, and Raed al Attar, a top Hamas commander, came out of hiding for a specially arranged summit. To this day it is unclear if this was a reckless decision caused by the shock of Deif's rumored assassination or a swift attempt by the two to scoop up their commander's authority and replace him before someone else did.

The first tip about the summit came two days after the attempt on Deif's life. The Shabak had picked up Abu Shamala and al Attar's tracks in the southern town of Rafah, a known Hamas stronghold that straddles Gaza's border with Egypt. It was a race against time. Every minute counted. Any reckless movement in the area might cause the senior Hamas commanders to flee, and Israel would lose its chance to strike another blow at Hamas's top echelons of power.

Israeli drones hovered over the area, trying to understand what was happening. The information flowed into the Shabak's command center. Once Israel received final confirmation that the two senior Hamas figures were inside the building, it took less than a minute for the missile to be launched. Later, when the rubble was cleared, the Palestinians admitted that the two arch-terrorists had been killed. Israel had just dealt Hamas another deadly blow.

Abu Shamala and al Attar were born a few months apart in 1974. They hailed from the Rafah refugee camp, one of the most crowded parts of the world—so crowded that IDF soldiers were afraid, already in those days, to venture inside. The two were raised as staunch Islamists and from an early age received a daily dose of Israel-hatred in their neighborhood mosque. At the age of 17, they started their careers in Hamas's military wing as guards of sensitive Hamas facilities. Soon enough, they gained their commanders' trust and were being trained for terror operations. They made a name for themselves in the 1990s, after carrying out a series of shooting attacks. In one incident, in 1994, they shot and killed IDF captain Guy Ovadia from the Nahal Brigade. Afterward, they carried out another attack in the area of the Kissufim Checkpoint—the main entryway from Israel into the Gaza Strip—and killed a 17-year-old aspiring Israeli air force pilot.

The Shabak succeeded in tracing the attacks to the duo, but then they disappeared. It was as if they had fallen off the face of the earth. Following the strict "Wanted Code" of Palestinian fugitives, Abu Shamala and al Attar regularly swapped safe houses and identities, avoiding contact with family members and friends. They walked the streets in disguise, rarely asking for assistance from fellow Hamas terror operatives. They trusted only themselves.

In 1995 it seemed that the duo's careers had come to an end. They were caught by the Palestinian security forces on suspicion of murdering a Palestinian Authority security officer in Gaza. These were the days after Yasser Arafat had returned to Gaza, and his security forces were trying to exert control. But then, Abu Shamala and al Attar were released, symptomatic of the Palestinian "revolving door," which often saw terrorists rounded up, imprisoned and eventually set free. Upon release, Abu Shamala joined the Palestinian Authority security forces, but after a few months, longing for his childhood friend, he ditched his uniform and returned to Hamas. To prove his renewed loyalty, he murdered another Palestinian Authority security officer. The two were again captured by the Palestinian Authority and this time were sentenced to life in prison. With the outbreak of the Second Intifada, in 2000, though, the two were again released alongside other high-risk prisoners to join the armed struggle against Israel.

Learning of their release, the Shabak renewed its manhunt. A decade passed. Abu Shamala advanced to the rank of Hamas's Southern Division commander, with responsibility for the Rafah Brigade, led by his old friend Raed al Attar. The Shabak succeeded in locating the pair's hiding place on several occasions, but each time, they slipped away.

Over the years, Abu Shamala and al Attar were involved in the planning and execution of dozens of terror attacks against Israel, including a number of attacks using tunnels that crossed under the border and into Israel. In 2002, al Attar helped plan an attack on an Israeli military outpost near the Kerem Shalom border crossing, which killed four IDF soldiers. In 2004, al Attar's men dug a tunnel under another IDF outpost, filled it with explosives and blew it up, killing six soldiers. In the summer of 2006, they were both involved in the infiltration of Hamas terrorists into Israel via a cross-border tunnel and the ensuing abduction of IDF soldier Gilad Shalit. Shalit, who remained in Hamas captivity for five years, was finally released in 2011 in a swap for more than 1,000 Palestinian security prisoners.

In addition to digging tunnels, al Attar was behind the establishment of the Nukhba—Arabic for "selected ones"—an elite Hamas force trained to fight and maneuver through the tunnels on foot and motorbikes. At the beginning of Operation Protective Edge, in the summer of 2014, al Attar personally oversaw the infiltration of 13 Hamas terrorists into Israel via a terror tunnel. Later, a force from his brigade was involved in an attack in Rafah, during which Hamas abducted the body of an IDF officer.

With the targeted killing of Abu Shamala and al Attar, it seemed that a power vacuum had again been created in Hamas. But it wouldn't last long. A few months after the operation ended, the IDF revealed that Mohammed Deif had survived the attempt on his life. Apparently a number of the bombs dropped that night had failed to explode. Deif was injured but alive. The hunt would continue.

✳ ✳ ✳

In its nearly seven decades of statehood, Israel has become the first country to master the art of targeted killings, integrating it into regular military doctrine and operations. It is a tactic Israel has successfully used on battlefields for two decades and is a story that combines cutting-edge technology, high-quality intelligence and Israel's best and brightest minds.

According to a 2010 United Nations report, a targeted killing is a premeditated act of lethal force employed by countries to eliminate specific individuals outside their custody. The particular act of force can vary from a drone strike to a cruise missile or a special forces raid.[1]

Targeted killings weren't invented by the State of Israel. They were put into use in the biblical era, during the Roman rule over the Land of Israel, through the Ottoman period and from the beginning of the Zionist settlement of what was then known as Palestine. Underground Jewish militias—Haganah, Etzel and Lechi—employed targeted killing tactics against their opponents.

After the establishment of Israel, the state continued to carry out targeted killings and assassinations. In the 1950s, Israel killed two Egyptian intelligence officers who had helped Fedayeen militants launch a series of attacks against Israeli towns and communities. In the 1960s, Israel sent letter bombs to German scientists who were developing missiles for Egypt. In 1972, following the murder of 11 Israeli athletes at the Munich Olympics, Prime Minister Golda Meir authorized the targeting of anyone discovered to have taken part in the attack. The retaliation for Munich was supposed to be the last occasion on which Israel officially killed people out of vengeance. Afterward, the policy changed; Israel would now target someone only to prevent future attacks from happening.

"It's not an eye for an eye," a former head of Shabak said about the new policy. "It's having him for lunch before he has you for dinner."[2]

In 1988, Palestinian terrorist Abu Jihad, responsible for

numerous attacks against Israel, was killed by an elite Israeli hit team in Tunisia, and in 1992, an Israeli Air Force combat helicopter used a Hellfire missile to kill Hezbollah leader Abbas Musawi in southern Lebanon. Both were senior terror leaders responsible for numerous attacks against Israel and in the midst of planning many more.

After the signing of the Oslo Accords between Israel and the Palestinian Authority, in 1993, the use of targeted killings dramatically decreased as both sides tried to give peace a chance. The killings did not completely stop though. In 1995, Fathi Shaqaqi, leader of the Islamic Jihad terrorist group, was gunned down on the streets of Malta, and a year later, Yahya Ayyash, Hamas's top bomb maker, nicknamed "The Engineer," was killed when a mobile phone filled with explosives blew up next to his head. Alongside the successes, there have also been failures, the most prominent of which occurred in 1997, when Mossad agents were captured in Jordan trying to spray a deadly poison into Hamas leader Khaled Mashal's ear.

Almost all of these operations were attributed to Israel, which refrained from taking responsibility. The idea was to target a select few terrorists but to deter the many who would know that Israel has the ability to reach them, no matter where they might be.

But then, in late 2000, everything changed. The Second Intifada erupted, and Israel found itself facing an unprecedented wave of terror, supported by Yasser Arafat's Palestinian Authority. Israel was up against a well-armed Palestinian force that had given suicide bombing attacks the regularity and efficiency of an assembly line.

In one incident, undercover IDF soldiers shot and killed a senior Tanzim terrorist next to his house in Jenin. A few weeks later, another terrorist was killed when his mobile phone exploded. The heads of the Palestinian terror groups understood that Israel was returning to its post-Munich assassination policy. This suspicion was confirmed in November 2000, when Israel conducted its first publicly acknowledged targeted killing near Bethlehem. An Israeli Apache helicopter fired a laser-guided missile at a car, killing senior Tanzim leader Hussein Abayat. A couple of months later, Masoud Iyyad, an officer with Force 17—a Palestinian commando unit run by Arafat—was killed in another helicopter strike. Israel claimed that he was working to establish a Hezbollah cell in the Gaza Strip.

The use of attack helicopters, particularly in the West Bank, represented a dramatic escalation by Israel. Whenever the assassinations were done by aircraft, Israel claimed responsibility.

At the same time that the killings by Shabak and IDF were piling up, so were the number of suicide attacks inside Israel.

A soldier from an elite IDF reconnaissance unit on assignment in the West Bank. IDF

Unlike the First Intifada, during which the ratio of Palestinians killed to Jews killed was roughly 25 to 1, now it was suddenly 3 to 1.[3]

By mid-2001, Palestinians had succeeded in carrying out dozens of suicide attacks against public buses, bustling coffee shops and packed discotheques. The Intifada wasn't going away, and pressure was mounting on Israeli leaders to take more aggressive action. Something needed to be done to stop the terror wave.

But not every terrorist could be apprehended, especially those operating deep inside Gaza. So IDF commanders came up with the idea of streamlining targeted killings—to penetrate enemy ranks and eliminate terror leaders. There was no time to waste. Legal guidelines were quickly drafted, and rough tactical standards were approved. The pace quickly picked up with the full support of the public; one newspaper poll conducted in July 2001 found that 90 percent of Israelis supported the tactic.

One IDF chief of staff carried a pad around with him with hundreds of names of wanted men. Sometimes the list reached 1,000. The targets came from the list. Each terror group—Hamas, Islamic Jihad and Tanzim—had its own color. When a target was hit, the name was crossed off with an X.

In July 2002, though, the support started to wane. Salah Shehadeh, head of Hamas's military wing, was at the top of Israel's most-wanted list. Shehadeh was one of the driving forces behind Hamas, its ideology and its operations. He was directly involved in the planning and execution of deadly terror attacks against Israelis, but because he was in Gaza and frequently switched residences, an arrest operation was virtually impossible.

An airstrike was approved, and on July 22, an F-16 fighter jet dropped a one-ton bomb on a house in Gaza City where Shehadeh was staying. In addition to Shehadeh and an assistant, 13 civilians—including women and children—were killed.

The international community erupted in protest, accusing

Israel of violating international law with the disproportionate killing of civilians. An Israeli human rights NGO petitioned the country's Supreme Court, and, under pressure, the government decided to establish a special commission to probe the validity of the strike.

While the Supreme Court, in a landmark 2006 decision, ultimately legalized targeted killings, the IDF understood that it could not continue the same tactics of dropping one-ton bombs or Hellfire missiles to eliminate terrorists holed up with civilians. It needed to develop more precise weaponry and also put in place strict and clear tactical procedures that would keep collateral damage to a minimum.

One missile developed at the time came with a warhead that included a mere 200 grams of explosives, which could, without harming bystanders, blow up a single apartment in a high-rise building or a car or motorbike traveling down a busy road.

Intelligence-gathering methods also underwent modifications, with tighter control over the decision-making process that led to a targeted killing. Drone use, in turn, saw a dramatic increase, since a targeted killing was rarely conducted without a drone first surveying the target.

But weapons and intelligence were not enough. The IDF still faced tremendous difficulty reaching terrorists embedded within civilian infrastructure—in hospitals, mosques or even private homes. One key test of a targeted killing is whether the results of the strike will be disproportionate to the killing of a single individual. In other words, if a wanted terrorist is hiding in a hospital, the bombing of the building would obviously be disproportionate to the value of killing a single terrorist. If he is hiding in a home, though, with one or two civilians, the decision could be different.

To attempt to deal with this challenge, in January 2009, the

IDF improvised and, on the battlefield, developed a new tactic called "knocking on the roof." The IDF was a few days into Operation Cast Lead, the first large-scale anti-Hamas operation in Gaza since Israel's unilateral pullout some three and a half years earlier. Intelligence, collected diligently over the previous year, indicated that a large number of homes were being used to hide arms caches. But the IDF knew it couldn't just bomb homes even if they had been transformed into legitimate military targets. Ahead of the operation, the IDF and Shabak collected phone numbers of the homes so they could call and warn the residents to leave before the bombings. The system proved extremely effective the first 54 times. On the 55th attempt, it failed.

That day, after the phone call was made, the residents climbed to the roof of the home and just stood there, waving at the IDF drone hovering above. Back at IDF headquarters, a debate broke out about what to do, and the strike was canceled.

The next day, the IDF called another home, and the same thing happened. "We understood that we had lost the tactical advantage," one IDF officer who was in the war room that day recalled.

This posed a serious dilemma. If the house wasn't bombed, the rockets it was hiding in the basement could be used the next day in attacks against Israel. On the other hand, there were women and children in the home. Israel couldn't just bomb it.

But then, a few officers in the Southern Command came up with a new idea: call the home, wait for the residents to climb to the roof, and then order a nearby attack helicopter to fire a small missile at a corner of the rooftop. The missile they were thinking of using had a small warhead with low shrapnel dispersion so that, if it was fired correctly, no one would be injured.

In one of the first uses of the new tactic, the IDF ran its usual playbook and called all of the phones in a three-story building that, according to Israeli intelligence, was hiding a large Hamas weapons cache. On the line were IDF officers, who, speaking Arabic, urged the residents to leave the building immediately, before it was bombed. The residents were undeterred. They climbed to the roof and waved at the drone they couldn't see but knew was above. The message was clear: they had no intention of leaving.

A nearby attack helicopter received authorization to fire a burst of machine-gun fire into an adjacent empty field. Some of the rooftop residents took the hint and fled the building. But some youngsters, realizing that their home would be destroyed, remained defiant. They were staying on the roof.

The pilot then received authorization to fire the missile at a corner of the roof. When it struck, the remaining people on the roof thought Israel was destroying the building even though they were still there. They fled, leaving an empty home for the air force to bomb. In a video of the attack, the building collapses, setting off a series of secondary explosions from the large weapons bunker hidden beneath. Israel was literally knocking on Palestinian roofs.

The more the new tactic was used, the more the IDF saw a continuous drop in civilian deaths. In 2002, the civilian-combatant death ratio was 1:1, meaning that a Palestinian civilian was killed for every combatant killed by the IDF. By the beginning of 2009, the ratio had dropped to 1:30.

That drop was due in part to the IDF's unique tactics, but just as significant a factor was the number of accurate weapons and smart bombs it was using in airstrikes. During Operation Cast Lead, the IDF dropped more than 5,000 missiles in the Gaza Strip. Over 81 percent of them were smart bombs, an unprecedented percentage in modern warfare. In comparison, during the beginning of the Iraq War in 2003, coalition forces used smart bombs 68 percent of the time, and during the Kosovo War of 1999, the percentage was a mere 35 percent.[4]

With this drop in civilian deaths, the Israeli government's confidence in the IDF's operational capabilities increased. Thanks to the successful targeting of terrorists and minimizing of civilian casualties, international pressure on Israel also dropped.

The 9/11 attacks in America marked a turning point in the use of targeted killings. The US found itself at war against an enemy that, unlike a conventional military, used fighters dressed as civilians who hid among women and children. Senior US military delegations came to Israel and particularly to the Southern Command to learn from Israel's experiences in hunting terrorists in the Gaza Strip. They were interested not just in the tactics the IDF used but also in how the IDF connected all of the dots among the informants, the Shabak agents, high-tech surveillance, military intelligence analysts and the air force.

The Bush administration decided to adopt targeted killings in Afghanistan and Iraq, and upon becoming president, in 2009, Barack Obama expanded the use to an unprecedented number of terror organizations and countries. In 2016, the US military revealed that it too was using the "roof-knocking" tactic in airstrikes against ISIS targets in Iraq.

Targeted killings, which had originated in Israel, were becoming the global standard in the war on terror.

✳ ✳ ✳

But why Israel? How did this small country come to set the world standard for how to combat terror?

The answer, we believe, can be found among the occupants of two nondescript office buildings on different sides of Tel Aviv. On one side are the Shabak headquarters, home to some of the most elite field agents in Israel. On the other side are military intelligence headquarters, where all raw intelligence flows and is analyzed by young IDF soldiers and officers. These two buildings are where Israel's innovative tactics and high-tech weapons are fused with a talented pool of intelligence agents and analysts.

In terms of prestige, the Rakaz course is for the Shabak what the Air Force Flight Academy is for the IDF. Thousands of citizens, aged 25 to 30, compete for places in the course, which is held only once a year. A select few are accepted, and even fewer complete the training.

The course starts at the Ulpan, the Shabak's language school. In Israel's early years, most Shabak Rakazim were born in Arabic-speaking countries and moved to Israel among the various waves of immigration to the state. But for over two decades, non-Arabic-speaking candidates have arrived at the gates of the Ulpan, and emerged 42 weeks later fluent in Arabic. They are capable of speaking with businessmen, politicians and farmers. They correspond over the Internet in Arabic, know full verses of the Koran by heart and are intimately familiar with the customs and culture of the Palestinians of Hebron, which differ from those of the Palestinians of Jenin or Gaza.

After graduating from the Ulpan, Rakazim are sent to the Shabak Intelligence School for a 10-month course. This is the point when the Rakaz trainee leaves his identity behind and becomes part of Israel's anti-terror shadow war. On the same day, the trainee receives a nickname that will accompany him until his last day in the Shabak—the name he will use when recruiting informants—as well as the specific geographical area he will be responsible for.

The new Rakaz learns to respect Islam as the religion and culture of his enemy. His handshake will be genuine, and when he converses by phone with a source, he will know how to listen not only to the information he is being told, but also to his source's mood and tone of voice as well as to noises in the background. The Rakaz will know everything about his informants—if one of their sons failed his math test, the date of a spouse's birthday and what the local gossip is that week on the other side of the neighborhood.

A talented Rakaz deepens his acquaintance with the diverse demographic, socioeconomic, political and social makeup of the population in his area of operations. He gets to know the members of the different Hamulas (Arabic for "clans"), the NGOs, the layout of the streets where the affluent residents live, who got married yesterday and who is expected to receive a large inheritance when his sick father passes away.

The Rakaz is trained to be a skeptic. When he sees a woman in a hijab pass him on a street, he needs to look twice to make sure she is not a terror suspect Israel has spent years hunting. When shops on Shuhada Street in Hebron shut down for a strike, he will need to contemplate the possibility that someone is preparing for a terror attack there. When a few residents of the Balata refugee camp near Nablus buy canned goods, he needs to consider the possibility that they are hiding a wanted man or an abducted Israeli soldier. Every Shabak agent lives by the following principle: "Not everything you see is what it appears to be."

Being a Rakaz involves a constant battle of wits. The Rakaz obtains his information through personal connections and mutual trust. This is how generations of Rakazim have been taught. The work of a Rakaz is anything but simple. Should he create an intimate relationship with an informant or keep a distance so the informant retains respect for him? Is a Rakaz allowed to endanger an informant's life to obtain information? What should a Rakaz do if the head of a terror cell suspects one of his operatives is an Israeli agent and asks him to shoot an IDF soldier to prove his loyalty?

Former Shabak chief Yaakov Peri started his career as a Rakaz in the West Bank. He described the work as "the art of courting"

Palestinians to collaborate with Israel and to betray their family and friends. The best Shabak sources, he explained, were wooed not because of the benefits—the money, the medical care they could now provide their families in Israel or the trips overseas—or from pressure. Rather, they were simply taken in by the Rakaz's personal charm.

A Rakaz's work is dynamic. In the past, the structure of a terror cell was clear. A cell leader had a deputy and then the regular cell operatives beneath him. Nowadays, a terror cell can be operated from a headquarters overseas with a cell consisting of Palestinians from the West Bank who don't even know one another. In 2015, for example, the Shabak uncovered terror cells in the West Bank directed from Gaza, Qatar and even Turkey.

Each member constitutes one step on the way to an attack without ever meeting another member. The first cell member buys a car and parks it next to a building that has been rented by another member. The third member places the bomb inside the car while the fourth drives the car to the target and the fifth detonates the bomb with a phone call.

In Protective Edge—Israel's anti-Hamas operation in the Gaza Strip in the summer of 2014—Rakazim accompanied soldiers from the IDF's Nachal Brigade as they invaded the town of Beit Hanoun, in northern Gaza. The infantry officers were amazed by the Rakazim's scope of knowledge: their familiarity with the names of the streets and the residents, down to the smallest detail, like what was hiding behind a trapdoor in the kitchen of a senior Hamas activist. They knew all of this without having set foot before inside Gaza.

This familiarity and knowledge is obtained not just through informants but also by studying the Gaza landscape. One method is to train with the unique type of simulator used by the Israeli

Air Force, which allows pilots, before they engage in an actual mission, to fly virtual bombing runs over terrain identical to that of a future target.

In the past, the threat profile from the West Bank and the Gaza Strip was of shooting attacks, stonings, Molotov cocktails and the occasional explosives laboratory, where pipe and roadside bombs were assembled. In recent years, the Shabak Rakaz has needed to focus also on terror tunnels, the development, production and smuggling of missiles from Iran and Hamas's potential use of drones. One moment, a Rakaz can be speaking to a Palestinian businessman about taxation and the cumbersome import-export process in Gaza, and a few minutes later, he might be engaged in conversation with a tunnel digger to find out the size and make of a new rocket recently smuggled into Gaza.

The work is a battle of minds—a constant attempt to be one step ahead of the enemy.

On the other side of Tel Aviv, in another unimpressive office building, are the headquarters of the Military Intelligence Directorate's Research Division. This is the place to which all intelligence flows. The division's job is to sift through the tremendous amounts of data accumulated by all of the country's different sensors—spies, satellites, drones, media and more—analyze it all and then predict what will happen. Will Iran violate the nuclear accord it reached in 2015 with Western superpowers led by the United States, known as the P5+1? Will Mahmoud Abbas agree to an unconditional renewal of peace talks with Israel? And how stable is King Abdullah's regime in Jordan?

The Research Division is meant to provide Israel's military and political leadership with the best-educated guesses in response to these questions. It is a tricky business with many moving pieces.

In mid-August 2014, a group of analysts gathered for a crucial meeting. Forty days had passed since Israel launched Protective Edge. Israel needed a way to shorten the operation and end the fighting, which would ultimately carry on for another week and become Israel's longest war since 1948. The intelligence officers entered the room after leaving their mobile phones in a special brown box outside. Iran, Hezbollah and Hamas were doing everything possible to eavesdrop on Israel. Even a deactivated cell phone could be used as a listening device.

The soldiers had nicknamed the room in which they were sitting "Shatzi," an acronym for some of their names. Accessible only to officials with the highest security clearance, the room is entered through a thick steel door opened with an electronic keypad. Closed-circuit cameras keep a watchful eye. A Hezbollah flag with a dedication to an officer who recently retired hangs on one of its walls, a constant reminder of the enemies they are fighting. Inside is another door opened with another keypad that accesses a small war room. Here are TV screens, computers and encrypted phone lines with immediate access connecting to other relevant IDF and Shabak offices. Soldiers man their stations 24/7, watching the screens, analyzing images and instructing reconnaissance drone operators on what clues to look for as they search for targets. It is a Sisyphean task but one that, if done right, has a potentially amazing payoff.

The Shatzi team is an elite squad within military intelligence. During Israeli operations in Gaza, for example, the team members prepare kill lists and hunt their targets, the men whom Israel is trying to eliminate. To be able to do that, Shatzi teams learn everything they can about the targets—their routines, where they live, where they go and whom they might be sleeping with on the side. The information is constantly updated so that when the strike order comes through, they will know how to find their targets.

Back at the August 2014 meeting, the intelligence officers raised a number of proposals for how to hurt Hamas and get the organization to stop its rocket attacks. One need quickly became clear: Israel had to find a way to eliminate senior Hamas operatives. The IDF knew the locations of many of them, but the problem was that they were hiding in hospitals and mosques or surrounding themselves with civilians. The potential collateral damage—particularly to civilians—meant that most potential strikes were nonstarters.

The meeting adjourned with new and clear instructions: work around the clock to collect information on high-value targets and expand the available strike options. At the same time, the analysts needed to be prepared to receive authorization from the cabinet to take out a senior Hamas operative. It could come anytime.

Some days passed. The Shatzi room was a constant buzz of activity. It was Friday, and Lieutenant S.—IDF regulations forbid the publication of an intelligence officer's full name—reluctantly took a much-needed break after 11 consecutive weekends on duty. It didn't last long. A few hours later, his Mountain Rose—the bulky, military-issued encrypted Motorola cell phone—rang. "We have a green light," his commander said. "We are going for al-Ghoul."

Moments earlier, the security cabinet had approved the targeted killing of Mohammed al-Ghoul, Hamas's money man, responsible for transferring millions of dollars to the organization's military wing and for financing the construction of cross-border terror tunnels into Israel. The information in his head was even more valuable than the stacks of dollars and shekels he carried around with him. He knew the bank account numbers, the names of the trusted money exchangers in Cairo and Amman and the location of smuggling tunnels along the Egyptian border that could still be used to get money into Gaza.

Inside the Shatzi, the computer screens were displaying real-time footage from drones hovering above al-Ghoul's house. He had been holed up inside his home with his wife and three children, but fresh intelligence indicated that his family was heading to the home of his in-laws. Al-Ghoul would be left alone. That was the moment Israel was waiting for. A car pulled up, and al-Ghoul's wife and children got in. Within a minute, though,

A photo released by Israel in 2008 of a Hamas military training camp in the Gaza Strip.   IDF

al-Ghoul also left the house, got into his car and started driving. The drone above confirmed that he was alone.

The echoes of Israeli Air Force bombings in the area did not frighten al-Ghoul as he left his house. The Shatzi officers had already analyzed the roads he could potentially take and identified a number of ideal strike points. The objective was to choose a place where it would be possible to kill al-Ghoul and no one else. At one point, the defense minister and IDF chief of staff called the war room for an update. By now, authority over the operation had moved to the senior air force officer on shift. Only he could give the final attack order.

The drone followed al-Ghoul for some time. He stopped at another home and pulled out what appeared to be a bag of cash for a group of Hamas operatives. While the green light had been given to strike, the air force commander was still waiting. He wanted a sterile attack without dead civilians. When the missile was finally launched, everyone held their breath. The explosion ripped apart al-Ghoul's silver sedan, killing him instantly. Dollar bills swirled up in the air, scattering throughout the street.

Lieutenant S. started his military service in the air force, as a commander of new recruits. He climbed the ranks and was appointed platoon commander. His personality and leadership skills stood out; his name came before the air force top brass, which was searching for a group of creative officers and brilliant minds to run its intelligence desk, the place the air force studies strike options, builds target banks and thinks up innovative ways to hunt Israel's most wanted.

After three years of training—including some of the most rigorous security and personality tests—S. was appointed to a role in an intelligence squad responsible for tracking senior terrorists in the Gaza Strip. He had a knack for the intelligence work, piecing bits of information together and accurately predicting how his targets would behave as they tried to outsmart Israel. After a few months, he was given command over two research teams. Together with a group of colleagues, S., who is in his twenties, developed a unique research tool used in military intelligence.

S. is convinced that the young age of IDF soldiers in critical positions like his is a definite advantage over Israel's intelligence counterparts around the world.

"We are more vigorous in our desire to change. We always talk about how to innovate and come up with new ways and ideas to reinvent the rules," he said. "This is a definite advantage."

In Israel, young intelligence analysts like S. have direct access to the top military and political echelons. They have to grow on the job and think on their toes. The big difference is that for S. and his fellow team members, the battlefield is not an ocean away.

"What I do is for my own benefit—protecting my family, my friends and my nation," he said. "I see the results of what I do on a daily basis, and if not today, then I will see them in the next war."

S. is just one example of a pool of IDF intelligence analysts making life-and-death decisions. He tracks wanted terrorists and watches regular-looking apartment buildings, beneath which Hamas or Hezbollah are suspected of storing rockets and arms. If one of his targets moves, he is updated by the operations desk. If renovations start on a building he is watching, he marks it down. Everything is noted. Everything is viewed suspiciously.

S. and his fellow analysts receive the daily routines of their targets and know if they are going to visit their mistresses or mothers. If they are going to the supermarket or the laundromat. These analysts have the authority to sound an alarm and convene the top military brass just because they noticed something out of the ordinary. This type of responsibility among such young officers is unprecedented in other Western militaries. In Israel, there are hundreds of such officers.

Over the years, Israel's targeted-killing apparatus has reportedly been adapted to all of the country's different fronts and "areas of interest." The challenge is immense and targets change daily. With the Middle East in the throes of a historic upheaval, borders change as well. In Syria, for example, this has been particularly difficult. Until the civil war erupted there in 2011, Israeli Intelligence needed to track a state and a military with a clear hierarchy. By 2016, though, in Syria there were Hezbollah fighters, Iranian Revolutionary Guards, Al Qaeda cells and thousands of ISIS fighters roaming the country. One Syrian city could be controlled by Bashar al-Assad and another, just a few miles away, by the Islamic State.

The world has not stood by and applauded Israel's use of targeted killings. Inside various international institutions like the United Nations, Israel is often accused of war crimes, crimes against humanity and violations of international law because of its use of targeted killings. An ongoing legal campaign has forced Israel to create significant legal oversight for how it acts and whom it decides to act against.

As nation-states and conventional militaries surrounding Israel and in the wider Middle East continue to disintegrate, the necessary act of striking at an adversary's power base is becoming extremely complex. Terror groups like Hamas, Hezbollah or ISIS don't have clear power sources. They don't usually have territories, let alone clearly defined bases. What they do have, though, are leaders—commanders and fighters whose deaths can sometimes amount to deadly blows to their continued operations. It will likely be the Shatzi's job to make sure that happens.

7

CYBER VIRUSES

About 200 miles south of the Iranian capital of Tehran is the small and ancient town of Natanz. It is a place known for its cool climate, quality homegrown fruit, terra-cotta brick buildings and some nearby mystical Sufi shrines.

On August 14, 2002, this small town made its way onto the front pages of every major newspaper in the world. The National Council of Resistance of Iran (NCRI), an Iranian opposition group, held a press conference at the Willard Hotel in Washington, DC, and revealed the existence of an underground and heavily fortified uranium enrichment facility that had been built near Natanz.

The news was startling. The world knew about Iran's construction of a nuclear reactor at Bushehr—started by the Germans and continued by the Russians—but Tehran always claimed that its program was for peaceful purposes. Rumors had circulated for years about secret sites, but none were ever discovered. Finding them in Iran's vast desert wilderness was in any case believed to be almost impossible.

But on this humid day in DC, the NCRI revealed the existence not just of Natanz but also of a heavy water reactor being built near Arak that would, in a few years, be able to produce weapons-grade plutonium.

Natanz, though, was the focus. It wasn't just any facility. Heavily fortified, it had been built underground. Each of its halls was built 70 feet deep, protected by thick concrete as well as another layer of steel to prevent penetration by air-to-surface missiles. An ordinary airstrike would not be effective. The Iranians had clearly learned the lessons of Israel's previous strike against Iraq's nuclear reactor in 1981.

Analysts immediately suspected that the Mossad was behind the information revealed that day by NCRI. The source, though, didn't make a difference. Iran had been caught red-handed building an off-the-books nuclear facility. The world could no longer ignore what was happening in Iran. The ayatollahs were building a bomb.

Israel used the revelation to stress the danger Iran posed to the world. The pressure bore fruit, and the International Atomic Energy Agency (IAEA)—the UN's nuclear watchdog—was eventually granted access to Natanz. Its inspectors confirmed the installation of advanced centrifuges built according to specs from blueprints obtained in Pakistan.

Gulf states, afraid of Iran's nuclear program and regional ambitions, made sure that the story stayed on the world's agenda. Israel did its part as well, repeating the mantra that all options, including military ones, were on the table. The Iranian Revolutionary Guards took those threats seriously and bolstered defenses around the facility with anti-aircraft guns and SAMs.

Natanz had a clear purpose. Its two underground halls—each about 100,000 square feet in size—were built to hold tens of thousands of centrifuges, the big steel machines that are used to enrich uranium. If Iran wanted a nuclear weapon, it was going to enrich the uranium there.

In 2009 something mysterious started happening at Natanz. The centrifuges were breaking down. One day, it would be a bunch at one end of a hall. The next day, it would be a few on another side. There seemed to be no correlation between the malfunctions and the computers that controlled the cascades of centrifuges and showed that everything was normal. It was a mystery.

It would take some time, but in November 2010, the IAEA confirmed that Iran had suspended work at Natanz. A few days later, realizing the secret was out, Iranian president Mahmoud Ahmadinejad downplayed the extent of the damage but admitted that Iran's enemies had succeeded in inflicting a limited amount of damage on computers at a number of nuclear facilities. The source of the problem, he claimed, had been identified and contained.

He was lying. More than 1,000 out of the nearly 9,000 centrifuges at Natanz had been decommissioned. Someone had succeeded in knocking out over 10 percent of Iran's centrifuges without firing a single bullet.

The question was: How?

The answer was called Olympic Games, a secret operation reportedly launched in 2006 by Israel and the US. These Olympics, though, had nothing to do with sports but were a covert, highly classified plan to develop cyber weapons that could be used to knock out Iran's nuclear program.[1]

For both countries, the idea was new, and its chances of success were unknown. Nevertheless, it was extremely attractive. A cyber weapon would not leave fingerprints, would be untraceable and, most importantly, would not require fighter jets to fly from Israel or the US to bomb Iran. On the other hand, the attack would have to be as devastating as a missile strike.

The Israel–US collaboration came at a time of heightened Israeli-American defense ties. Israel was still traumatized from the outcome of the Second Lebanon War, and the US was looking for ways to assuage Jerusalem's concerns over Iran. If the joint development of a cyber weapon could postpone an Israeli strike, then from the White House's perspective, it was worth pursuing.

The chosen target was Natanz and, more specifically, the Siemens industrial computer systems that control the rows of centrifuges there.

The worm, which would later receive the name Stuxnet, targeted the device that controls the speed of the centrifuge's motor, used to spin around and enrich uranium. The Stuxnet code changed the frequencies of the converter, first to higher than 1,400 hertz and then down to 2 hertz—speeding it up and then nearly halting it—before settling at a little more than 1,000 hertz.

Essentially, Stuxnet caused the engines in Iran's centrifuges to increase and decrease their speed, just enough so they would eventually break.

At first, it was difficult for the Iranians to believe that the problems at Natanz were caused by a virus, since the facility's computers were not connected to the Internet. As time passed, though, the only realistic possibility was that one of the facility's computers had been infected with a virus by a covert agent, likely using a portable flash drive.

To succeed, it would not have been enough to insert the worm into the Natanz computer system. The attacker would have had to know the exact layout of the Iranian computer system down to individual wires so the worm could know how to travel and jump from one system to the next. Israel and the US would have needed to obtain these exact blueprints.

Next, they would have had to find someone whose computer they could infect and use as a springboard to jump into the Natanz network. For this, the CIA reportedly relied on Israel, which it believed had informants deep inside Iran's nuclear facilities.[2]

Such techniques have worked in the past. In 2008, an American soldier found a few memory sticks scattered on the ground near a military base in the Middle East. The sticks had been deliberately infected with a computer worm, and the foreign intelligence agency behind the operation was counting on a soldier's human instinct to pick up a stick and insert it in his computer to see what was on it. The result was nearly devastating, delivering a worm into the computer system of the US military's Central Command that took 14 months to eradicate.[3]

Stuxnet was reportedly created by teams working in Israel and the US. In Israel, Unit 8200—the IDF equivalent of the US National Security Agency (NSA)—led the work, with assistance from the Mossad.[4] In the US, the lead contractor was the NSA. Unit 8200 and the NSA have worked together for years.

Ralph Langner, a German IT expert, was one of the first independent experts to analyze the Stuxnet code after it leaked out of Iran. His mouth dropped open as he stared at its 15,000 lines of code. In his analysis, the worm had succeeded in setting back Iran's nuclear program by two years.[5]

"From a military perspective, this was a huge success," he said.

The virus, Langner and other experts discovered, had been sitting in Iranian computers for a number of years and had successfully spread to over a dozen countries. Around 100,000 computers were found infected, of which about 80,000 were in Iran.

The "sabotage" that Ahmadinejad spoke about, though, was something far greater. Stuxnet was not just a virus. It was a weapon that dictated new rules. Its use changed modern warfare.

Computer experts immediately assumed Israel was behind the worm. Israel had the greatest interest in stopping Iran's nuclear program and was its most vocal critic. There were also some clues discovered in the code: a number that apparently referred to the date of the assassination of an Iranian-Jewish philanthropist and the word "Myrtus," a possible reference to Esther, the Jewish queen from ancient Persia, whose heroism is celebrated on the holiday of Purim.[6]

While Israel never confirmed or denied its involvement, there was one big clue. In 2011 the IDF chief of staff was retiring, and, as is customary, the military put together a farewell party for him. Israel's leadership was present—the prime minister, the president, the defense minister and, of course, the top military brass. During the event, a movie made as a tribute to the IDF commander's career was shown. There was his participation in the First Lebanon War in 1982 and various operations he had commanded over in the Gaza Strip and the West Bank. Toward the end, footage related to another operation attributed to Israel flashed briefly across the screen—Stuxnet.

On the surface, it is strange that Israel and Iran are enemies. The countries do not share a border, they are the only two non-Arab states in the Middle East, and they actually formalized diplomatic ties shortly after Israel was established, in 1948. The Islamic Revolution of 1979 changed all that. From the closest of friends—rumors at the time were that Israel was mulling the sale of advanced ballistic missiles to Iran—the countries turned into the worst of enemies.

In the beginning, the conflict was mostly about Iran's support of anti-Israel terror groups—Hezbollah in Lebanon as well as Hamas and Islamic Jihad in the Gaza Strip. By the late 1990s, Israeli intelligence was indisputable—Iran was moving full-throttle ahead toward obtaining a nuclear weapon.

Iran's nuclear program had started decades earlier, with US assistance. In the initial days after the revolution, the mullahs suspended all nuclear work, but it didn't take long for them to realize the potential of what they had: nuclear power could turn Iran into a global superpower.

In 2002, Israel stepped up its fight against Iran's nuclear program. That year, Prime Minister Ariel Sharon appointed Meir Dagan, a veteran IDF general, as head of the Mossad, Israel's shadowy international spy agency.

Dagan had made a name for himself throughout decades of service in the IDF as a daring soldier and innovative tactician. His bravery was renowned; he earned a Medal of Valor for grabbing a grenade from the hands of an enemy fighter.

In 1970, Sharon, then head of the IDF's Southern Command, chose Dagan to command a unit called Sayeret Rimon, whose soldiers, disguised as Palestinians, raided the Gaza Strip to hunt down PLO terrorists.

The tone for Dagan's tenure as Mossad director was set by the way he decorated his office. On the wall of the modest room was a black and white picture of an old bearded Jewish man, wearing a tallit—a Jewish prayer shawl—as he knelt down in front of two Nazi soldiers, one with a stick in his hand, the other carrying a rifle.

"Look at this picture," Dagan would tell guests. "This man, kneeling down before the Nazis, was my grandfather just before he was murdered. I look at this picture every day and promise that the Holocaust will never happen again."

The Mossad's work reportedly paid off. Iranian scientists began to disappear. Some defected to the West. Others were assassinated by masked gunmen on the streets of Tehran.

In January 2007, for example, Ardeshir Hosseinpour, a senior Iranian scientist, was found dead in his office at the Isfahan conversion plant, a key facility in Iran's nuclear program. He was believed to have been "gassed" to death. In January 2010, Masoud Ali Mohammadi, another key Iranian nuclear scientist, was killed when a motorcycle loaded with explosives blew up outside the front door of his Tehran home. A few months later, in November, a bomb went off in downtown Tehran, killing Majid Shahriari, another top nuclear scientist.

Assassinations were not the only instrument used to undermine Iran's nuclear program. There have also been countless reports of sabotage. In 2007, for example, electrical components used to regulate voltage currents at Natanz mysteriously exploded, destroying dozens of centrifuges. In other cases, spy agencies recruited companies around the world to purposely sell Iran faulty hardware for its nuclear program and missile program.

While the Mossad succeeded in slowing Iran's progress, the ayatollahs did not give up. They diverted funds from welfare projects, from the nation's medical institutions and from universities. Scientists received bodyguards and were banned from leaving the country. Security was tight everywhere.

By the mid-2000s, Israeli intelligence had concluded that Iran had mastered all of the technology needed to make a bomb. Now all that was left was the decision to do it. If Israel didn't come up— soon—with a way to stop that from happening, then a military strike, Israel's last resort, would become the only option. The war that would ensue was spoken about in catastrophic terms.

Stuxnet wasn't the first cyber attack the world had seen,

but its effectiveness highlighted the world's deepest fears. Until then, attacks against Internet sites had caused limited damage. In 2007, for example, an attack against Estonia temporarily paralyzed banks, government ministries and local media. Stuxnet, though, represented much more. It had the potential to bring national railways to a stop or shut down a city's power grid. As Dmitry Rogozin, Russia's ambassador to NATO at the time, said: Stuxnet had the potential to "lead to a new Chernobyl."

Iran decided to fight back and, within a year of Stuxnet's discovery, established its own cyber unit, investing over $1 billion to create potent offensive and defensive capabilities.

Iranian computer experts realized that Stuxnet was just the tip of the iceberg and that it could be followed by attacks of a more sophisticated and devastating nature. Indeed, in 2012, Iran discovered another virus—called Flame—that had infected computers throughout the country. Flame, which was reportedly developed by Israel and the US, was designed to map out computer networks and steal data from infected computers.[7]

In response, Iran acquired new tools for operating on the Dark Net, the shadowy side of the Internet, where hostile elements and cyber criminals can be found. It also is believed to have shared its knowledge and technology with Hezbollah—Iran's proxy in Lebanon—a force multiplier in the event of a war one day with Israel.

That same year, US intelligence revealed that Iran's cyber capabilities had increased dramatically in "depth and complexity." In the two years after Stuxnet, Iran stepped up its cyber attacks, hitting the Saudi state oil company, a Qatari natural gas firm and a number of US banks.[8]

Concerned with this growing threat, Prime Minister Benjamin

Netanyahu summoned his military secretary for a special meeting to brainstorm about Israel's next steps. While the country had built some formidable defenses, Netanyahu was worried that Iran would be able to penetrate them, considering its growing cyber capabilities. His military secretary suggested consulting with Professor Yitzchak Ben-Israel, a former colleague of his from the air force and a world-renowned expert on national security, technology and, particularly, cyber warfare.

The meeting was scheduled for a few days later, and Ben-Israel got the impression from Netanyahu that time was short and that action was needed to prevent and defend against a potential Iranian attack. The fact that hackers from the former Soviet Union were offering cyber weapons to the highest bidder did not help. Israel was on edge.

Ben-Israel was an example of the Israeli success story. Born in Tel Aviv in 1949 to parents who fought with the Lehi, one of the pre-state Jewish underground organizations, Ben-Israel learned at a young age that Israel could not take its security for granted.

After high school, he enlisted in the IDF as an academic, studied physics and math and quickly made a name for himself as an innovative thinker, serving in a number of key operations and intelligence roles in the air force. In 1972, Ben-Israel received the prestigious Israel Defense Prize for the development of a new bombing system for Phantom fighter jets, significantly upgrading their attack capabilities. A year later, he stood by the side of the IAF top brass during its darkest hour, the Yom Kippur War, when more than 100 of its combat aircraft were shot out of the sky. He was part of a key team of engineers and technical experts who helped the IAF rehabilitate itself after the war, steering it in the direction of technological prominence.

In the winter of 1992, Ben-Israel happened to be in Berlin for meetings with the German military. During one meeting, his

German colleagues started talking about the Internet and its potential military applications. Ben-Israel wasn't sure what they were talking about, and the German officers offered to send him some academic papers by email. That also didn't mean much to the senior Israeli officer. Only later that year would the Internet, developed originally as a US military project, penetrate the civilian market and reach Israel.

When he returned to Tel Aviv, Ben-Israel immersed himself in the works of Alvin Toffler, an American writer and futurist known for books including *Future Shock*, *The Third Wave* and *Revolutionary Wealth*. Ben-Israel was fascinated by Toffler's ideas about what the world would look like when the Internet took over. His head spun with ideas of what the Internet would mean for the IDF and the future of warfare.

By 1995, Ben-Israel felt the time had come for the IDF to establish a computer warfare unit. At the time, no one even used the world "cyber." He met with the IDF chief of staff, who, to Ben-Israel's surprise, approved the idea and allocated a massive budget. That was until the air force got wind of the plans. "It's too early to invest in such technology," the air force officers protested out of a desire to retain their status as the military's top technological unit.

Despite the opposition, Ben-Israel managed to find an ally in a young colonel from Aman named Pinchas Buchris. A decorated officer and commander from the elite Sayeret Matkal, Buchris was known in military circles for his role in various operations that included the 1976 mission to free Air France hostages in Entebbe, Uganda. The operation was one of Israel's most daring and demonstrated for the whole world the IDF's far reach.

Buchris would go on to command Unit 8200 and serve as director general of the Defense Ministry. But then, as a young colonel, he decided to establish a small subunit in Aman to oversee computer warfare. Ben-Israel and Buchris agreed that the unit would be dedicated to offensive and defensive computer-based operations. Not many people even knew what that meant at the time, but soon enough, ideas—such as using computers to attack enemies—started to emerge.

In 1998, Ben-Israel was promoted to the rank of major general and made head of Mafat, the Defense Ministry's R&D directorate. His responsibilities were diverse. On the one hand, Mafat needs to accompany and provide support for ongoing military operations, but on the other it also needs to keep an eye on the future and invent new weapons that Israel will need to win the wars to come. Cyber warfare soon became one of Ben-Israel's focuses.

In 2000, Ben-Israel wrote a letter to Prime Minister Ehud Barak warning that Israel was vulnerable to a computer attack. If a hostile force uncovered the ideas under development in the IDF's computer unit, he told Barak, the country could be brought to a standstill. The potential damage was extensive.

Barak—who was busy at the time trying to reach a peace deal with Yasser Arafat and stop a new Palestinian uprising—took Ben-Israel's warning seriously. He instructed the National Security Council to review the vulnerability of Israel's critical infrastructure—water, electricity and gas. It would take some time, but two years later, the government officially established a new unit called Re'em—a Hebrew acronym for National Information Security Authority—tasked with protecting the nation's infrastructure.

When he met Netanyahu, in 2011, Ben-Israel received a green light to set up several working groups to think up ways Israel could prepare for the war to come, a war that would undoubtedly include cyber attacks. He recruited 80 experts from different fields, including academics and senior government officials, whom he then split into different committees, each engaged in putting together a list of recommendations, pertaining to the economy, the high-tech industry, the military and even Israeli universities. Buchris was chosen to head the eighth and secret committee, which dealt with the defense establishment. Within a few months, Ben-Israel had returned to the prime minister with a 250-page report and a long list of recommendations.

Even then, it was clear that the cyber field was crowded and that a multitude of players—the Mossad, Aman and Shabak—were already engaging in cyber activities without any coordination. A common language among all of the different agencies needed to be created. To this end, Ben-Israel recommended establishing the National Cyber Bureau under the auspices of the Prime Minister's Office, a single and central authority responsible for coordinating among all of the different players and capable of coming up with quick solutions for threats and problems. Members of the various committees recommended strengthening cyber studies at Israel's universities, allocating additional government budgets to bolster the private industry and establishing new cyber-research centers. By 2016, there were five such centers in Israel, one of which—the Blavatnik Interdisciplinary Cyber Research Center at Tel Aviv University—was headed by Ben-Israel.

The secret committee, headed by Buchris, recommended the establishment of a military cyber command within Unit 8200. The largest unit in the IDF, 8200 is responsible for gathering and processing signal intelligence—basically anything that is transmitted by phone or over the Internet.

Unit 8200 is synonymous with Israel's amazing high-tech boom in the past 20 years. Its graduates have gone on to establish some of Israel's most successful tech companies, making it one of the most sought-after assignments in the IDF. Service in the unit ingrains its soldiers with cutting-edge technological skills alongside a keen sense of entrepreneurship and innovation.

The unit's soldiers work hand in hand with Israel's growing list of cyber companies and in many cases develop technology in-house. Their job is simple to explain but difficult to execute— listen to conversations throughout the Arab world, intercept emails and track current events as they unfold.

While Israel officially stays silent on its cyber capabilities, there is no hiding the fact that it is today a global leader in cyber security, exporting more than $6 billion a year in cyber products, rivaling Israel's annual defense exports. With just eight million people, Israel has captured about 10 percent of the global cyber market—hundreds of high-tech companies have been established by Aman graduates alone—putting it on the same level as countries like the United States, China and Russia.[9]

But how did Israel achieve such cyber superiority? According to Ben-Israel, the state's standing as a cyber superpower can be traced back to three decisions made by David Ben-Gurion when Israel was established in 1948.

As we showed in Chapter 1, it was clear to Ben-Gurion that the State of Israel would not be able to defeat its Arab enemies based on numbers alone. Instead, if the state was to survive, three essential goals had to be met. First, Israel needed to establish a "people's army" and draft an unprecedented percentage of both male and female citizens. That is how, when statehood was declared, the IDF became a military that, at any given time, included about 5 percent of the country's entire population. In Western countries, the percentage of the overall population serving in the military ranges from 0.2 to 0.4.

Ben-Gurion's second decision was to ensure that the military put an emphasis on quality and not just quantity in its recruitment. He pushed for an educated military, made up of intelligent and innovative soldiers who knew how to draw on the Jewish tradition of education and scholarship imported to the fledgling state by the various waves of immigrants from Europe and North Africa.

"This is not a matter of genetics but of culture," Ben-Israel told us when we met at Tel Aviv University, citing the fact that over 20 percent of Nobel Prize winners are Jewish, including a handful of Israelis.

The third decision by Ben-Gurion was to promote the importance of science and technology within the army. It was no coincidence that when the IDF was formally established, it was the only military at the time to set up a Scientific Corps in addition to the traditional infantry, navy and air force.

One question arose, though: if soldiers were drafted at 18, where would the military get its engineers, mathematicians and physicists? That's why the IDF established the Atuda Program, a special academic track for Israel's youth. The soldiers would first study in university and then be drafted for six years, three more than mandatory. It was a big ask, but one that Israeli youth immediately accepted.

Atuda was viewed within the military as an elite cadre of high-quality human capital. Its graduates—known as the IDF's "wonder kids"—served in the military's different technical units, developing and operating its most advanced systems. Some of the graduates signed on for service tracks beyond the six years, completing their military service at the age of 28 or 30 with significant experience. Many returned to university to become academics, but most joined Israel's burgeoning tech industry.

After the Yom Kippur War, in 1973, the Atuda program rose

in importance. The French embargo—imposed in 1967 after the Six Day War—was still in place, and Israel was left without spare parts for its planes, tanks and other weapons systems. This huge vacuum led Israel to the realization that it needed to expand the domestic defense industry even if just to make ends meet.

Over the next three decades, the Israeli military industry grew from just a few thousand workers to more than 40,000 employees. This came at a heavy cost to the state. The steady growth of the different companies forced successive governments to cover their annual deficits. The companies strived to develop and produce the most advanced technology and weapons systems, but fear of losing the technological advantage stopped them from exporting their products to foreign militaries.

While aware of the potential security risks, Israel decided to slowly open its defense companies to the world. By the mid-1990s, Israel's annual defense exports exceeded $1 billion, and within a decade they reached $4 billion. The process advanced slowly, with calculated risks. But Israel was on a clear track to becoming a military superpower.

✳ ✳ ✳

One of the Atuda "wonder kids" is Yaniv Harel, who until 2015 served as the head of the Defense Ministry's Cyber Department. The youngest child of Holocaust survivors, Harel grew up in the suburbs of Haifa, in northern Israel, and in 1992 enlisted in Atuda and studied electronic engineering at the Technion. Toward the end of his studies, he applied to join one of Aman's classified intelligence units. His high grades opened doors for him, and he quickly stood out as an unconventional engineer who did not shy away from breaking down bureaucratic barriers between different intelligence agencies. For Harel, partnerships were empowering, a belief that went against the conventional culture within the intelligence community, where most commanders preferred to keep sensitive information close to their chests. For one of the classified projects Harel managed, he was awarded the Israel Defense Prize.

After 15 years of service, in 2007 Harel hung up his uniform to go back to school and pursue a doctorate in strategic management at Tel Aviv University. He had mixed feelings, since he knew that he would not be returning to the unit where he had grown as an officer.

After completing his degree, Harel passed up a senior position back at Aman to join the Defense Ministry, where he became head of Mafat's Cyber Department. He found the army in the midst of an intense debate about how big its cyber unit needed to be. One school of thought advocated for the addition of hundreds of new soldiers and officers to unit 8200 to develop new systems and weapons. Harel disagreed. He pushed the unit to open up to the civilian market and establish stronger ties with Israeli companies that were making amazing technological advances.

"We don't need more people to duplicate what is already being done," Harel told his fellow officers. "We just need better relations with civilian companies."

What Harel was suggesting was revolutionary. Israel's cyber capabilities were a closely guarded secret and until then had been developed mostly in-house, by units like 8200.

Harel understood, though, that the global cyber train had already left the station and was moving fast. If the IDF didn't pick up its pace and expand its development and production capabilities, it would fall behind its enemies. In his first year in office, Harel spearheaded 15 new partnerships between the military and Israeli start-ups. By 2014, before he stepped down, there were 80 such projects.

Harel attributes Israel's success in becoming a cyber superpower to the country's culture of accepting failure. "There are other cultures in which an individual is personally eliminated if he fails, and this makes him more rigid in his thinking, since he is worried that action will endanger his position," Harel told us from his office at the headquarters of EMC, where he heads up the company's Cyber Solutions Group. "The average Israeli dares to take risks without fear of failure."

Israel's other advantage, Harel said, can be seen in the IDF, where junior officers are encouraged to raise their hands in critical discussions and openly disagree with their superior commanders. This culture aims to prevent a scenario in which critical information remains in the hands of a single soldier who, for fear of its impact on the military hierarchy, prefers to keep it to himself.

Hanging on one wall in Harel's office is a picture of the Wright brothers next to one of their first airplanes. For Harel, the picture is a symbol, not of the past, but of the future. When the Wright brothers flew for the first time, they didn't know that they were opening the door to a new dimension of warfare, one that would lead to the establishment of air forces, stealth fighter jets and attack drones. The door to the new dimension of cyberspace is already open. Where it leads is still unclear.

✳ ✳ ✳

One legendary operation demonstrates this new type of warfare— one that is still spoken about in hushed voices and saw the combination of espionage, cyber warfare, electronic warfare and nuclear weapons.[10]

The first report that something had happened appeared in Syria's official news agency, SANA, on September 6, 2007. The agency claimed that Syria's air defense units had identified an infiltration by Israeli Air Force jets the night before, fired missiles at the jets and scared the pilots away. The report claimed that the jets had dropped their missiles in the desert without hitting any targets.

While news of the air raid was peculiar, it was not the first time Israeli jets had raided Syria. In 2003, four F-16s buzzed Syrian president Bashar al-Assad's summer residence in Latakia, where he was then vacationing, in retaliation for the killing of a young Israeli boy by Hezbollah rocket fire. Israel wanted to humiliate Assad and send him a message to rein in his Lebanese terror proxy. The planes flew so low that they apparently shattered some of the palace windows. A couple of months later, the IAF bombed an Islamic Jihad training base in Syria in response to a suicide bombing that killed 19 people. And then, in 2006, Israeli fighter jets again buzzed the Latakia residence to remind Assad that he would pay a price for giving refuge to Hamas's leadership in Damascus.

But this time, Israel was silent. Media requests for comment went unanswered. A couple of days later, the US State Department acknowledged that it had heard about the incident from secondhand reports but denied knowing anything more. The first proof that something had even happened appeared in the Turkish newspaper *Hürriyet*, which published photos of two Israeli fuel tanks discovered on the Turkish side of the border with Syria, the warplanes' apparent flight path. After a few more days, reports started to surface that the target was in fact a nuclear reactor Syria was building along the Euphrates River in northeast Syria.

The news was shocking. The world knew about Assad's support for Hezbollah as well as his large chemical weapons arsenal, but no one thought that he would try to build an illicit nuclear bomb. Behind the scenes, the IAEA started pressuring Syria to let its inspectors visit the site, but Assad refused, claiming that the attacked site was an empty warehouse.

The attack was reportedly carried out by 10 F-15 fighter jets. Minutes into the flight, the final attack order came through, and seven of the planes broke off from the formation, increasing their speed as they entered Syrian airspace. Seconds later they had already dropped their first bombs on a radar installation. Another two minutes passed, and the planes were over the nuclear reactor. They dropped their AGM-65 bombs, each one weighing about half a ton.

As the pilots began making their way out of enemy airspace, the Syrian military finally realized it was under attack and launched a salvo of anti-aircraft missiles blindly into the sky. The planes, by then, were long gone.

The pilots who flew that night into Syria learned of the true nature of their target only hours before. Until then, they had trained for a bombing mission in Syria but had not been told what the target would be. If true, it would be the second time Israel had destroyed a nuclear reactor, only this time, unlike the 1981 bombing of the Osirak reactor in Iraq, the pilots wouldn't be able to speak about it, either with their friends or with their families.

What is less known about the bombing of the reactor was Israel's reported use of an innovative electronic warfare cyber attack that tricked Syria's air defense systems, first making it seem that no jets were in the sky and then, in an instant, making the radar indicate that there were hundreds.[11]

The technology was revolutionary. The world had known about cyber attacks and electronic warfare. It had not yet seen the two used tactically as one.

It appeared that Israel had mastered a technology known in the US as Suter, a system that basically fools radar systems, making them see something that does not exist. It had been developed by the Pentagon several years earlier but was not believed to be within Israel's reach. How Israel got the technology remained a secret, but it appeared to have been developed in-house, by Israeli engineers.[12]

In the end though, it really didn't make a difference. While most of the world focused on the destruction of the nuclear reactor, the attack that night was transformative—Israel had reportedly used cyber technology on the battlefield.

✳ ✳ ✳

When Bashar al-Assad was appointed president of Syria in the summer of 2000 to replace his father, there was hope in the West that the British-educated ophthalmologist would open the country to the world and initiate extensive reforms and possibly even a peace deal with Israel. Assad, it turned out, had other plans. He strengthened ties with Hezbollah and Iran and established a strategic relationship with North Korea.

In 2004, the NSA detected an increasing number of phone calls being made between North Korea and Syria. On the Syrian side, many of the calls appeared to originate in a place in the Syrian desert, near the town of Dir a-Zur, along the banks of the Euphrates River. The NSA reportedly passed the information on to Unit 8200, its Israeli counterpart, which in turn established a team of analysts to try to figure out what was happening north of Israel.

At first, the possibility that Syria was cooperating with North Korea on a nuclear project seemed impossible. Construction of a reactor would leave a string of clues that would have been detected by Israel. The working assumption was that the relationship centered on the development of ballistic missiles. Both countries had strong ballistic missile arsenals and were likely trying to improve their capabilities. It seemed unlikely that Assad would try to revive the idea of a Syrian nuclear weapon, something his father had contemplated but ultimately rejected in the 1990s, when he passed up the opportunity to purchase nuclear technology from rogue Pakistani scientist A. Q. Khan.

So in late 2006, the Mossad reportedly dispatched a couple of agents to London to consult with MI6, the British foreign espionage agency, and try to crack the Syrian–North Korean alliance.[13]

When the Mossad agents arrived in London, they had a surprise waiting for them. According to a report in *Der Spiegel*, a senior Syrian government official was visiting London at the same time and had checked into a luxury hotel in Kensington. The Syrian regularly went around London carrying a handbag with a laptop inside. On one occasion, though, when he exited the hotel for a meeting, he left the computer in his room. The Mossad team reportedly received permission to break in and hack the computer. The operation took a few seconds; the agents broke into the room, installed a Trojan horse and left.

Within minutes, the Trojan horse was transferring information from the Syrian's computer back to Mossad headquarters near Tel Aviv. The hard drive was an intelligence treasure trove, containing construction plans and photos of the reactor Syria was building. One of the pictures showed two men in their fifties, one an Asian in a blue tracksuit and the other an Arab. The men were Chon Chibu, a senior North Korean nuclear scientist, and Ibrahim Othman, head of the Syrian Atomic Energy Commission. The pictures changed everything. All of Israel's assumptions about what North Korea was doing in Syria had been wrong.[14]

But now Israel had to find the site. That job was given to 9900, Aman's Visint unit, responsible for collecting and processing all visual intelligence collected by the country's satellites. It didn't take long before the analysts zeroed in on the suspected location: a few low buildings surrounded by trees in northeast Syria. The engineers who chose the location had invested a lot of thought into ways to avoid attention as well as detection by reconnaissance satellites. The façade of the suspected reactor resembled one of the many byzantine fortresses that dot the Syrian countryside, and it was built in a ravine so it could be seen only from a high vantage point. There were also no security measures nearby—no air defense systems or anti-aircraft guns to raise suspicion. But there was one building they couldn't hide: the water-pumping station that any heavy-water reactor requires and its accompanying pipeline, which ran in the direction of the Euphrates River, less than a mile away.

Prime Minister Ehud Olmert reportedly received an immediate update. He gathered two special forums. The first included Defense Minister Amir Peretz; Mossad director Meir Dagan; the IDF chief of staff, Lieutenant General Gabi Ashkenazi; the head of Aman; and the Israeli Air Force commander. The second forum included three former prime ministers: Shimon Peres, Benjamin Netanyahu and Ehud Barak.[15]

The security forum put together three different attack options. The first was a quiet airstrike to be carried out by a small number of fighter jets, which would provide Israel with some deniability. The second option was a noisy air operation with a full show of force. The purpose would be to publicly humiliate Assad. The third option, and possibly the riskiest, was to send Israeli Special Forces into Syria to plant explosives and destroy the reactor.[16]

Olmert and Peretz ordered the IDF to prepare all three, although it was understood early on that the preference was for the quiet option that would not leave fingerprints.[17]

The clock was ticking. IDF intelligence analysts warned that the attack needed to take place before the reactor was activated. If not, an attack on an activated reactor would contaminate the Euphrates and potentially harm Syrian and Turkish civilians.

Even with the pictures and construction plans collected in London, Israeli intelligence agencies were still doing all they could to get more information. They wanted to be certain. In March 2007, they reportedly had another opportunity. Othman, the Syrian nuclear chief, was in Vienna for an IAEA meeting. Mossad agents broke into his house, installed a Trojan horse on his computer and left without a trace. The information again left no doubt—Syria was trying to build a nuclear weapon.[18]

Olmert decided that Syria needed to be stopped, but he didn't want to attack. He wanted the Americans to carry out the operation. In April, US secretary of defense Robert Gates was scheduled to visit Israel, the first visit by a Pentagon chief in nearly a decade. It was perfect timing. When Gates arrived, his staff was informed that Peretz would be coming by the hotel in a few hours for a private off-protocol meeting. It was a strange request, but the Israeli defense officials insisted. "He has an important message to convey," they told the Americans.

At the same time, Dagan, the Mossad chief, was making his way to Washington, DC, for a meeting with Vice President Dick Cheney and National Security Advisor Stephen Hadley. The idea was to present the intelligence collected on the reactor at around the same time.

The Americans were surprised by the revelations, but Olmert

An Israeli Air Force F-16 takes off from a base in southern Israel in 2014.   IDF

reportedly decided to go a step further and asked President George W. Bush straight out to attack the reactor. Bush mulled over the option, according to US officials familiar with the conversations, but ultimately turned down Israel's request. Already fighting two wars in the Middle East, Bush refused to open another front with another Arab country. If there was going to be an attack, it would have to be carried out by Israel.

The eventual bombing of the reactor demonstrated the new kind of warfare Israel was reportedly engaging in—a combination of spies, commandos, satellites and cyber warfare culminating in an airstrike.

What the future holds is unclear. There is no rulebook yet for how wars are fought in cyberspace. Does a cyber attack like Stuxnet or the installation of a Trojan horse on a foreign government computer constitute an act of war like an airstrike or a ground invasion? Stuxnet set back Iran's nuclear program, but Tehran never retaliated. Syria's nuclear reactor was destroyed, and it, too, remained quiet.

If a cyber attack is launched against Israel, what will the country do? Will it respond or remain quiet like Iran and Syria?

These questions have yet to be answered. What is clear is that future wars will look completely different from the wars of the past. Already today, militaries like the IDF have new roles for soldiers who serve as "cyber warriors," each armed with a keyboard that, with a few keystrokes, can potentially bring a country to its knees. Whatever Israel does, the world will be watching.

8

DIPLOMATIC ARMS

The unmarked Boeing 707 airliner took off from Ben-Gurion Airport and landed an hour later in the southern resort town of Eilat. It was late at night, and for the tourists walking the beach boardwalk, the plane was a familiar sight. After an hour on the ground, the plane took off again, this time eastward. Ten hours later, it landed in Kolkata, India, spent a couple of hours on the ground to refuel and took off once again. The plane's destination was the city of Guangzhou, capital of a southern Chinese province. There, a group of German-speaking Chinese navigators boarded the plane for its fourth and final journey—to a sealed-off military base on the outskirts of the Chinese capital.

After landing in Beijing, the "foreigners"—as the group was called—were finally allowed off the plane and taken to a nearby compound that had once served as the Belgian embassy in the Chinese capital. Venturing outside the compound, they were told, was out of the question.

The "foreigners" barely spoke to one another, assuming that

all of the compound's cabins were bugged and their conversations were being taped. If there was something important to discuss, they went outside, into the cold and polluted Chinese night.

The Chinese didn't really know who the two dozen men were. They had been told that the group consisted of foreign businessman who had connections with several leading international defense companies, including some from Israel. That was just the cover. In reality, the delegation was led by the CEO of Israel Aerospace Industries, the leading government-owned defense company, who was joined by senior representatives from the Foreign Ministry and the Defense Ministry.

It was February 1979, and Israeli defense officials had set foot in China for the first time.

A long time in planning, the trip was a closely guarded secret. Israel and China did not have diplomatic relations. Nobody—except for the members of the delegation, the prime minister, the defense minister and a handful of others—knew about the trip. If word got out, Israel knew that the Americans would be furious. On the other hand, the US was on the verge of announcing official diplomatic ties with China, and if there was a time to go out on a limb, this was it.

It was a match of mutual interests. China was in the middle of implementing far-reaching reforms following the Cultural Revolution and was opening up to the West. In Iran, Ayatollah Khomeini had just returned from exile in France, and the Shah's regime had collapsed, meaning that Israel was about to lose one of its primary arms customers. China could fill that vacuum.

The man who opened the door to China for Israel was Saul Eisenberg, a Jewish billionaire who had fled to Shanghai like 20,000 other Holocaust refugees during World War II. After the war, Eisenberg built a financial empire in the Far East, becoming one of the first Westerners to do business in China, Japan and

Korea. He used these ties to interest China in Israeli weaponry and even donated his private Boeing 707 to transport the Israeli delegation on that maiden flight to Beijing in 1979.[1]

Nevertheless, the Chinese were apprehensive. They did not want to aggravate their traditional and long-standing allies—the Soviet Union and the Arab bloc—by suddenly opening up to Israel. Both Israel and China needed to tread carefully.

Eisenberg was familiar with Israeli defense products from other deals he had helped broker throughout Asia. After a series of preliminary meetings with the Chinese, he returned to Israel with a shopping list—an unorganized mix of missiles, radars, artillery shells and armor—and urged that the government send a delegation.

In Israel, the decision to go to China was not made easily. Prime Minister Menachem Begin was in the loop from the beginning but deferred questions regarding the shopping list to Defense Minister Ezer Weizman, whom he ordered to personally approve what Israeli companies could and could not sell.[2]

The Chinese were told that the delegation was a group of Eisenberg's friends, who had ties in Israel and could obtain arms as needed. The Israelis were just as confused about whom they were dealing with.

"Were they engineers? Intelligence operatives? Military officers?" one member of the Israeli delegation recalled. "They all wore these 'Mao Suits' tunics. We had no way of knowing who we were even talking to."

For the week that the delegation spent in Beijing, they were not allowed contact with Israel. A mother of one of the participants passed away while he was in China, but there was no way to let him know. He heard about her death only when he was on the plane for the flight back home.

During their stay, the "foreigners" presented the Chinese with brochures of different weapons they claimed they could get from Israel. The Chinese were impressed but did not make any commitments. Additional trips followed, some of them made on Israeli Air Force planes, which had their blue Star of David insignias removed. By then, the Chinese knew they were working directly with the Israeli government. Once the shopping list was finalized, it was brought to Begin and Weizman for approval.

The negotiations were a clash of cultures. The Israelis wanted to sign a contract that could be used as a general framework and be applied automatically to future sales. They were hoping for a shopping bonanza. The problem was that the Chinese were not used to complicated contracts. At one point during the negotiations, for example, the Israelis insisted on including a force majeure clause in the contract. "What is that?" the Chinese asked. After the Israelis explained to them that it freed the sides of a breach of contract in the event of an unforeseen act of God, the Chinese answered plainly: "Well, there is no need for that since we don't believe in God."

It took a year, but after lengthy negotiations, the sides finally reached a framework agreement. The first shipment—of tank shells—arrived in 1981.

The relationship continued, and even as the ties grew warmer, the deals still had to be done in complete secrecy. The Chinese, for their part, refused to come to Israel but agreed to sign hundreds of millions of dollars' worth of contracts on the basis of a couple of photos of weapons and an occasional piece of video. They never saw real production lines. In the arms sales world, it was an unprecedented leap of faith.

In 1985, the veil of secrecy lifted just slightly. For the first time, the Chinese agreed to issue visas to nine executives from Israel's agriculture industry, including a government official from the Agriculture Ministry, to learn some of Israel's innovative farming techniques. That same year, Israel reopened its consulate in Hong Kong, which it had closed a decade before.[3] Israel pressed Beijing to establish official ties, but that would have to wait for the collapse of the Soviet Union and the Madrid Peace Conference in 1991. Only after the Arabs openly sat with Israel would the Chinese do so.

In 1992, diplomatic ties were finally established. Civilian trade skyrocketed, growing from under $100 million in 1992 to over $8 billion 20 years later, turning China into Israel's number-one trade partner in Asia. That secret arms trip in 1979 made this possible.

※ ※ ※

Since Israel's inception, defense ties and particularly arms sales have played a role beyond their clear economic purpose of bringing billions of dollars into the Israeli economy. Surrounded by hostile Arab states, Israel has leveraged its superior technology and military expertise to establish diplomatic ties with countries—like Russia, China, Singapore and India—that normally would have shied away from the Jewish State.

The interests have varied. Some countries viewed Israel's rise as a state with admiration and wanted to try to duplicate its success. Others faced similar threats from countries that, like those on Israel's borders, operated Soviet weaponry. They wanted to learn from Israel's experience, honed by years of conflict and war.

While the weapons sales helped establish diplomatic ties, as in the case of China, they also came at a price, often creating tension between Israel and its number-one ally, the United States.

This was the case with the Phalcon, which was at the center of possibly the most disastrous arms deals Israel ever entered into with a foreign country. The Phalcon was an airborne early warning command and control (AEWC&C) aircraft, and its sale to China was expected to reach some $2 billion, Israel's largest arms deal ever. Israel saw benefit beyond the economic gain. China had influence over some of Israel's enemies, and Jerusalem hoped that the deal—which would solidify Israel as China's number-one arms supplier—would create leverage over Beijing's diplomatic thinking.

Talks over the sale of the Phalcon started in the late 1980s, even before diplomatic relations were officially established between the countries. In 1993, a year after Israel opened its embassy in Beijing, Prime Minister Yitzhak Rabin became the first Israeli head of state to visit China, and the talks gained momentum. The Chinese had issued a global tender, and Israel submitted an official proposal, alongside Russia and the UK.

AEWC&C systems play a critical role on the modern battle-field, providing real-time intelligence and radar detection needed to achieve and retain aerial superiority. Israel's Phalcon, developed and manufactured by Elta, a subsidiary of state-owned IAI, is one of the most advanced AEWC&C systems in the world, capable of tracking dozens of targets simultaneously. China planned to use the planes to see what was happening along its sea borders and to project its power throughout Asia.

After his trip, Rabin asked the Defense Ministry to update the Pentagon about Israel's decision to compete for the Chinese tender. The request was natural. Israel had reached an understanding with the US several years earlier that it could sell weapons to China as long as they didn't include American technology.

While Israel made an offer to China to install the Phalcon's systems—a mix of radar and electronic intelligence systems—on a standard Boeing plane, the Chinese insisted on using the Ilyushin, a Russian transport aircraft. This complicated things. Israel had never bought military hardware from Russia, the main arms supplier of Israel's enemies. Russia had also lost out on the tender. The last thing it was likely to do was help Israel.

But Rabin wanted the deal. So in 1995, he asked Boris Yeltsin for the plane. The response was positive but noncommittal. Two years later, Prime Minister Benjamin Netanyahu traveled to Moscow and finalized the sale. So that the Chinese angle would stay hidden, reporters who accompanied the prime minister were told that the plane was for Israel.[4]

After Netanyahu's visit, Moshe Keret, the CEO of IAI, flew to Moscow to work out the technical details. He managed to negotiate the price down to $45 million. The problem, though, was getting the Russian engineers to understand the new design Israel required for the plane so it could accommodate the radar systems that would make it a Phalcon. This proved slightly complicated, since, while the parent company had offices in Moscow, the Ilyushin was assembled in Uzbekistan and the engines came from Ukraine.

Once the deal for the plane was finalized with Moscow, Israel immediately signed the Phalcon contract with China and received a down payment. Everything seemed to be on track.

Netanyahu believed in complete transparency, and after closing the deal for the plane, he updated President Bill Clinton about the progress. The Americans weren't thrilled but refrained from voicing serious opposition. But then, in 1999, everything changed. Ehud Barak was elected prime minister, and a few months later, the Ilyushin landed at Ben-Gurion Airport. Within days, a picture of the strange-looking plane with the massive dome appeared on the front pages of local newspapers. The China deal was no longer a secret.

Once everything was out in the open, Chinese defense minister Chi Haotian decided to visit Israel for an update. It was a historic visit, the first by a Chinese defense minister. What Israel didn't anticipate, though, was the anger pictures of Haotian inspecting an honor guard in Tel Aviv would raise in the US.

The influential *New York Times* columnist A. M. Rosenthal, for example, condemned the Barak government for allowing the visit of "one of the ranking Tiananmen killers." The Jewish State, he wrote, shouldn't sell a single pistol to China, let alone technology that would potentially assist Beijing one day in shooting down US aircraft.[5]

The criticism escalated, and Congress started paying attention to what was happening. Suddenly all hell broke loose.

In April 2000, Defense Secretary William Cohen publicly condemned the deal and warned that it would degrade America's ability to operate freely in Asia. In June, the U.S. House of Representatives passed a nonbinding resolution objecting to the sale. The chairman of the House Appropriations Subcommittee on Foreign Operations, Representative Sonny Callaghan, suggested cutting US military aid to Israel by $250 million, the same amount China had already paid for the plane.[6]

"What will Mrs. Goldberg in New York say when her son serving in the Pacific is shot down because of a Jewish AWACS?" officials in the Pentagon asked their Israeli counterparts.

Keret flew to DC to see if he could get one of the top lobby firms on K Street to take up the Phalcon case. One firm said it would look into it. The rest said nothing could be done. Keret returned to Israel and understood that the deal was dead in the water. All that was left was to publicly acknowledge the flop.

By the middle of 2000, Barak had completed an Israeli withdrawal from Lebanon after an 18-year presence there and was getting ready for the Camp David summit, at which he would eventually offer Yasser Arafat a comprehensive peace deal, only to be rejected. To make peace with the Palestinians, Israel would

Chinese Rear Admiral Yang Jun-Fei is welcomed by Israeli Navy Brigadier General Eli Sharvit, upon docking in Haifa Port in 2012.   IDF

need US support—to be more exact, financial support to upgrade the IDF's capabilities and evacuate settlements. The Phalcon needed to be sacrificed.

The cancellation of the Phalcon deal, and a subsequent crisis in 2004 over allegations that Israel was using American technology to upgrade drones it had sold China years earlier, infuriated the Pentagon. In response, Israel's involvement in the development of the F-35 Joint Strike Fighter, a stealth fighter jet Israel planned to purchase, was suspended. Contact between the Pentagon and Israel's Defense Ministry dropped to a bare minimum. Israel had to choose sides, and under immense pressure, it agreed to completely and indefinitely stop selling weaponry to China.

The Phalcon affair was a critical lesson for Israel. Arm sales can help open doors, but if they are not done carefully, they can also potentially close them, sometimes in more than one country.

Israel was now stuck with the plane and a massive penalty it needed to pay Beijing for breaching the contract. By then, IAI had almost completed installation of the radar and associated sub-systems. It desperately needed to find a new buyer.

Keret went to India and secured a deal—not just for one Phal-con but for three, at a whopping $1.1 billion. India was a strange choice as a customer. The nation had established full diplomatic relations with Israel less than a decade earlier, and at the time of the sale, not much was publicly known about the ties between the two countries; few knew that those ties included a defense component.

As with China, Israeli-Indian relations got their start before diplomatic ties were officially established. Back in the 1970s and 80s, Indian officers frequently visited Israel to ask for help in the war they were fighting in Kashmir against Pakistan. The Indians were interested in learning new tactics, with an emphasis on the emerg-ing use of electronic warfare, an Israeli specialty. The deals were tiny and were carried out mostly through third-party vendors.

But while India benefitted from its relationship with Israel, it did so privately, due to a combination of Cold War alignments, fear of alienating its large Muslim population and a need to main-tain ties with the Arab world.

In 1990, the relationship moved into high gear. David Ivry, the former air force commander and now director general of the De-fense Ministry, secretly flew to London for a meeting with Indian defense minister V. P. Singh. The war in Kashmir wasn't going away, and the Indians needed to upgrade their military, which consisted mostly of old Soviet platforms.

A few months later, Ivry dispatched a delegation of senior IDF officers and representatives of defense companies to New Delhi. He then flew there for a meeting with Prime Minister P. V. Narasimha Rao. The meeting was the first between a top Israeli defense official and an Indian head of state.

Ivry was accompanied to the meeting by his immediate counterpart, director general of the Indian Defense Ministry. He came bearing a letter from Yitzhak Rabin, Israel's defense minister at the time, in which Rabin welcomed the prospect of establishing defense ties with India. After introductions, Rao asked his staff to leave the room. He wanted a few minutes alone with Ivry to hear what Israel was offering and how deep a relationship the Jewish State was interested in forging with India.

Ivry gave Rao an overview of Israel's military capabilities, showed him the letter from Rabin and explained what would become essential to Israel's emerging relationship with India.

"We don't have any strings attached to our sales," Ivry said. "We are a small country. Superpowers attach diplomatic conditions to defense ties. We don't."

Rao liked what he heard. Before reaching out to Israel, India had tried purchasing weapons from the US but was told that a long list of diplomatic conditions—mostly related to human rights—would need to be met before the sales could be approved.

Rao called the director general of India's Defense Ministry back into the room and gave him the green light to sign $2 billion worth of defense deals with the State of Israel. The Israelis had a lot to learn about how to do business in India, but that business ultimately paid off, with cumulative sales topping $10 billion by 2014, making Israel the top arms supplier to India after Russia.[7] The arms sales have helped turn the tide in India's diplomatic view of the Israeli-Palestinian conflict. In 2015, for example, In-

dia surprisingly abstained when voting on three anti-Israel reso-
lutions in the United Nations.

Singapore is another country with which Israel established
diplomatic ties after first forging military ties.[8] Singapore has
evolved into one of Israel's key allies and, according to different
reports, is the funder behind the development of some of the IDF's
most advanced weapon systems. Today, Singapore's military is one
of the best-equipped in the world and receives 20 percent of the
island state's overall budget.

Ties between Singapore and Israel date back to 1965, when
Singapore achieved independence from the Federation of Malaysia. At the time, it had no military to speak of, and while the British were helping Malaysia establish an army, the same offer was
not extended to Singapore. The country needed help and needed
it desperately.

Singapore's new defense minister, Goh Keng Swee, secretly invited Mordechai Kidron, Israel's ambassador to Thailand, to the
island. Kidron, together with a Mossad representative, arrived
within a few days of independence with clear instructions from
Jerusalem: offer military assistance to the fledgling state. The offer was attractive. Like Singapore, Israel had gained independence
not that long ago and had succeeded in establishing a powerful
military in a short amount of time. The two countries had something else in common: they were small in size and surrounded by
hostile states with Egypt and Syria along Israel's borders and
Malaysia and China near Singapore.[9]

Lee Kuan Yew, Singapore's founding father and first prime
minister, knew and liked Kidron. They had met a few years earlier, when Kidron had come to ask to open an Israeli consulate in
Singapore while it was still part of Malaysia. While he had faith
in the Jewish State, Kuan Yew ordered Keng Swee to put the

Israeli option on hold until he heard back from India and Egypt, two countries he had already turned to for assistance. A few days later, though, Kuan Yew received disappointing news. India wished Singapore "sincere good wishes for happiness and prosperity" but ignored its request for military assistance. Egypt's response was similar; it extended recognition of Singapore as an independent state but ignored a request to send a naval advisor. Kuan Yew was especially disappointed with Egyptian president Gamal Abdel Nasser, whom he had thought of as a personal friend.[10]

Left with no alternative, Kuan Yew gave the green light to proceed with Israel but gave instructions to keep the news under wraps in order to avoid provoking Malaysia, a Muslim country. Three months later, a group of Israelis arrived in Singapore led by a colonel named Jack Elazari. To hide the Israelis' true identity, the Singaporeans referred to them as "Mexicans."

Before leaving for Singapore, Elazari met with Yitzhak Rabin, then IDF chief of staff. "Remember," Rabin said. "We are not going to turn Singapore into an Israeli colony. Your job is to teach them the military profession, get them to stand on their own two feet and to eventually run the military on their own."[11]

That is exactly what Israel did. Instead of holding large-scale training courses, they focused on small groups of Singaporeans who had a bit of military experience and trained them as commanders. They also got to work building bases, writing doctrine and creating the necessary structure between the military and the political echelon. In total, some 200 commanders were enlisted and trained.[12]

Kuan Yew wanted a people's army like Israel, one that was based on a compulsory draft. In the beginning, though, the Singaporean leader wanted the Israelis to draft only people who were unemployed, what he called "society's primitive people." Kuan Yew cited the Japanese as an example. Japan, he said, fought against the intellectual British military in World War II and won, proving that intellect was not an asset for a soldier. Soldiers, he said, didn't need to think much. They needed to know how to take orders.

Elazari and his team disagreed. They explained to Kuan Yew that the Japanese soldiers followed their orders and fought valiantly in World War II since they were dedicated to their emperor. "It's not a matter of education," the Israelis explained. "But of motivation." Kuan Yew was convinced.

While the Israeli delegation continued its work, Kidron returned to Singapore and demanded quid pro quo. It was time, he said, that Singapore recognize Israel and that the two states send ambassadors to one another. Kuan Yew said that that idea was a nonstarter. Establishing ties with Israel would upset the local Muslim population and anger Malaysia, which was clearly aligned with the Arab bloc.

In 1967, the Six Day War broke out. While Kuan Yew was relieved that Israel won, the war presented Singapore with a new dilemma. The UN was debating a resolution condemning Israel, and Kuan Yew knew that if Singapore supported it, Elazari and his team would abandon the country. If he abstained or voted against the resolution, though, the world would immediately know that there was something going on between Israel and Singapore. He was stuck.

After intense deliberation, the Singaporean leader decided to abstain, in effect admitting that the island country had a relationship with the Jewish State. Once that happened, there was no reason not to give Israel what it wanted. Israel was allowed to open a trade office in Singapore in October 1968, and six months later, an embassy.

* * *

China, India and Singapore are three examples of countries with which Israel used what we call "arms diplomacy" to establish diplomatic ties. What Ivry told Indian prime minister Rao in 1990 about not attaching strings to arms sales has been a key principle in this effort, enabling Israel to become a dominant player in markets where other Western countries cannot easily enter—places like Africa, Eastern Europe and Asia.

When Israel was first established, it did not have many real friends—especially not in its immediate vicinity, the Middle East.

A small military, the IDF was not, and still is not, a large enough customer to incentivize local companies to develop high-end weaponry. For that reason, the vast majority of products that Israeli defense companies manufacture are exported. This way, the companies can keep production lines open and prices down for the IDF.

At IAI, for example, 78 percent of sales in 2014 were to foreign customers. This is the same basic breakdown for all of the large defense companies in Israel. This is unlike any arrangement elsewhere in the world. In the US, foreign sales make up a significantly smaller portion of sales for defense conglomerates. International sales at Boeing's defense division represented about 35 percent of its business in 2014. At Lockheed Martin, that figure was only 20 percent.

Take the Popeye missile, one of the Israeli Air Force's most advanced weapons, as an example. Developed by state-owned Rafael Advanced Defense Systems, the Popeye can accurately hit targets through a window from over 60 miles away. It is one of the air force's most sophisticated standoff missiles. But how many can the Israeli Air Force order? To keep prices down, it needs to export the missile. That is why one of Israel's most sophisticated weapons—the type of technology a country would usually want to keep to itself—has been sold to the US, India, South Korea, Australia and Turkey.[13]

In short, if Israel doesn't export the Popeye, Rafael won't be able to afford to develop and produce it.

But this is not always simple. Controversial sales, like that of the Phalcon to China, can create tension and suspicion between friends. Allies need to be able to trust one another, and arms deals can undermine that trust.

In 2005, following the crisis over Israel's China sales, the Defense Ministry established a new mechanism to oversee its defense exports. Until then, all defense sales were overseen by a Defense Ministry unit called SIBAT, whose primary task was to promote defense sales overseas and help companies make connections with foreign governments. Under the new mechanism, Israel established the Defense Export Control Agency (DECA), which registers exporters, approves sales and issues licenses to sell in foreign markets.

Israel also agreed to increase coordination with the United States. Basically, whenever a sensitive arms sale is on the table, Israel consults the Pentagon. Loss of independence was a heavy price but one worth paying to stay on Washington's good side.

That new mechanism was what brought a group of senior Israeli defense officials to the Pentagon in late 2008. They had come to discuss something unprecedented—a billion-dollar deal to sell Israeli military drones to Russia.

This was not a regular sale. Israel is the world's largest exporter of drones; it has sold unmanned aircraft to countries in Africa, Europe, South America, the US and Asia. But it has never sold drones—or any weaponry, for that matter—to Russia. For decades, Russia has served as the main arms supplier of Israel's enemies, particularly Iran and Syria. Russian arms, such as anti-tank missiles, have also found their way to Hezbollah and Hamas. If Israel sold its own technology to Russia, there was a strong likelihood it would one day find its way to Lebanon, Syria and Gaza.

Russia's interest in Israeli drones was sparked during the war it fought with Georgia in South Ossetia in the summer of 2008. The war lasted five days, and while Russia ultimately won—ending the war by recognizing the independence of South Ossetia and Abkhazia—the fighting exposed a severe decline in the Russian military's technological capabilities, particularly when it came to drones.

In the weeks leading up to the war, and amid growing concern that Russia was going to annex the breakaway territories, Georgia began flying drones on routine reconnaissance missions over the conflict zone. These weren't just any drones. They were Hermes 450s, manufactured by Elbit in Israel and used by the Israeli Air Force. In the span of three months, Russia shot down three drones. In one memorable downing, Georgia released a video showing a MiG fighter jet firing a missile at the drone and scoring a direct hit.

While the downings of the drones were impressive on their own, Georgia's use of drones highlighted a problem on the Russia side. To begin with, Russian drones were late to the battlefield and failed to provide real-time intelligence, forcing Moscow instead to dispatch fighter jets and long-range bombers for standard recon missions. One drone used during the war was the old Tipchak, which Russia later admitted made too much noise, making it easy to detect and intercept.

On the other hand, the Georgian military effectively gathered intelligence, largely due to its small fleet of Israeli drones.[14]

Weeks after the war ended, Russia turned to Israel and asked to purchase the Hermes 450, the same drone used by Georgia.

Israel was initially shocked. Russia had never before purchased weapons from a foreign country, let alone from Israel. But the war was a wake-up call for Moscow, which was willing to admit that it needed technological assistance.

Everyone agreed that no matter what, Israel could not sell drones that were still in operational IAF use. During the Second Lebanon War, in 2006, Hezbollah fired dozens of Russian anti-tank missiles at Israeli tanks. The last thing Israel could afford was to have its own drones one day be used against it.

But then defense officials came up with an idea. What if, by selling drones to Russia, Israel could prevent the sale of sophisticated arms that were supposed to be delivered to Iran or Syria? That would not only make it possible to live with the risk the drone sale posed, it could even make it worthwhile.[15]

In Israel, opinions were split. The Foreign Ministry supported the sale and claimed that it could help strengthen ties with Moscow, especially at a time when Iran was moving ahead with its nuclear program. The sale of drones, these officials argued, would provide Israel with real leverage over Russian policy on issues such as Iran. While the Defense Ministry was in favor of obtaining some leverage over Moscow, it had difficulty overcoming the genuine concern that the drone technology would one day find its way to Iran, Syria and then even Hezbollah in Lebanon and Hamas in the Gaza Strip.

At the time, there was one Russian arms deal that everyone in Israel agreed needed to be stopped at all costs: the delivery of the advanced S-300 air defense system to Iran. The original $800 million deal had been signed secretly in 2005, but under pressure from Israel and the US, Russia was delaying delivery.

Israel's reasons to even consider such a quid pro quo were simple. The S-300 was one of the most advanced air defense systems in the world, was combat proven, could track up to 100 targets simultaneously and had the potential to make an Israeli air strike against Iran's nuclear facilities impossible.

The Russians were well aware of Israel's concern regarding the S-300. It came up in almost every conversation. About a week after the war ended in South Ossetia, Israeli prime minister Ehud Olmert spoke by phone with Russian president Dmitry Medvedev. Russia was upset with Israel for supplying Georgia with arms and drones. During the conversation, Olmert agreed to a moratorium on Israeli arms sales to Georgia but also pressed Moscow on its sale of weapons to Syria and Iran.[16]

Officially, the Kremlin gave Israel assurances that it would not transfer weapons to Iran that could destabilize the region, a message that could be interpreted as a decision not to supply the S-300. At the same time, though, Moscow explained to Israel that if Iran met its obligations to the IAEA—the United Nation's nuclear watchdog—delivery of the S-300 would be reexamined positively.[17] Anyhow, the Kremlin argued, the S-300 was a defensive system, and Israel, if concerned about it, should simply not attack.

Russia refused to reveal its true intentions. In early 2009, for example, US senator Carl Levin visited Russia. Levin was chairman at the time of the Senate Armed Services Committee and had come to Moscow to try to increase cooperation on missile defense in the face of Iran's continued pursuit of a nuclear weapon. A vocal supporter of Israel, Levin also raised the S-300 sale and urged Deputy Foreign Minister Sergei Ryabkov to hold back from delivering the weapons system to Iran. But Ryabkov stood strong, saying that while the deal was currently frozen, it didn't help that everyone kept talking about it.

"The less we hear from Washington about this, the better," he said.[18]

News of the freeze did not alleviate Israeli concerns. In Jerusalem, some thought that an attack on Iran would need to be moved up so it could take place before the S-300 arrived.

Israel made sure to get this message out to some of the moderate Arab states it was friendly with in the Persian Gulf. United Arab Emirates (UAE) chief of staff Hamid Thani al Rumaithi, for example, met with Richard Olson, the US ambassador to Abu Dhabi, in early 2009 with an urgent request: that the US immediately deploy five Patriot missile defense batteries in the UAE. The reason was fear that due to the S-300 deal, Israel was on the verge of attacking Iran, and Iran would then retaliate against the UAE.[19]

"I need to be open and frank with you, there are changes in the region that concern us," Rumaithi told Olson. The Patriot batteries, he explained, would be deployed in and around Abu Dhabi to protect against potential Iranian missile attacks in retaliation to an Israeli strike. When pressed on what might precipitate an Israeli attack, Rumaithi referred to the delivery of the S-300 system. "I don't trust the Russians, I've never trusted the Russians or the Iranians," he added.

Back in Israel, the drone deal suddenly became even more urgent. The final decision, though, wasn't just in the hands of the Defense Ministry. If the Foreign Ministry vetoed the deal, the Defense Ministry could still bring the sale to the Israeli Security Cabinet, which had the authority to overturn the decision.

The Security Cabinet convened a number of times during 2009 to discuss the proposed deal. Russia wanted to purchase long-endurance drones like the ones Georgia had used during the war. Israel made a counter offer: it would consider selling drones, but only older models like the Searcher, which the air force had retired several years before.

In June 2009, Israel's new foreign minister, Avigdor Lieberman, flew to Moscow. That occurred during a period of flourishing Israeli-Russian ties, cultivated mostly by the Moldovan-born Lieberman. By that summer, five Israeli cabinet ministers had visited Moscow, tourism was at an all-time high, a free trade agreement was in the works and Russia was talking to Israel about hosting a Middle East peace conference in Moscow.

In some Washington circles there was concern that Israel was looking to replace the US as its primary ally. Israeli-US ties were frayed, in any case. Benjamin Netanyahu had been reelected as Israel's prime minister and was already knocking heads with Barack Obama, the new US president.

During his meetings in Moscow, Lieberman raised the S-300 sale. The Russians, who openly opposed an Israeli strike against Iran's nuclear facilities, told him that the S-300 was "only destabilizing if you are planning to attack Iran" and refused to rule out supplying the system.[20]

Lieberman walked away with a clear conclusion—that finalizing the S-300 deal was a matter of Russian prestige and would ultimately be carried out. If there was any doubt about the need to move ahead with the drone sale, it disappeared when Lieberman returned to Jerusalem.

A few weeks later, top American and Israeli officials gathered in Tel Aviv for the annual Strategic Dialogue, a forum established to discuss regional developments as well as practical ways to ensure Israel's qualitative military edge over its neighbors. The S-300 came up, and the Israelis revealed to the Americans their most recent discovery, that Russia was planning to move ahead with the deal if the United States continued its plan to deploy missile defense systems in Poland and the Czech Republic.[21]

If Israel was going to move ahead with the drone deal, now was the time.

Before it could sign with Moscow, Israel had to pass one more major hurdle, which was the United States. Russia and America were old adversaries, and Washington would not be happy with Israel's selling advanced drones to a country that once was—and in some circles still is—an enemy.

One of the first American officials to hear about the proposed sale was Mary Beth Long, the assistant secretary of defense for international security affairs. Long, who was the first American official to give a boost to the Iron Dome rocket defense system, was taken aback by the Israeli request mostly because it came at a time when Israel was asking the US not to sell advanced weaponry to Saudi Arabia and other Gulf States. "You are saying I can't have arms sales to Arab countries without approval and you are selling advanced weapons that may or may not be a derivative of US tech to Russia," Long told her Israeli counterparts.[22]

The Israelis argued. First, they claimed that the technology was not based on anything that originated in the US. Second, they told Long that the drones they would sell to Russia were just a slight qualitative improvement over Moscow's current capability and were no different from drones China was offering to sell the Russians.[23]

Long asked about the possibility that Russia would one day transfer the technology to Hezbollah or Iran, but the Israelis assured her that even if that happened, the drones currently in Israeli use were a generation ahead of the ones it would be selling Moscow.

If Israel sold the drones to Russia, Russia would occasionally need Israel for routine maintenance, spare parts and know-how. This dependence, Israel figured, could be leveraged into real influence over Russian foreign policy and, particularly, arms sales to Israel's Arab adversaries. If the Kremlin did something against Israeli interests, Jerusalem could ground the Russian drones by refusing to sell spare parts or perform routine maintenance.

Long brought the proposal to her superiors. A Pentagon team was established to review the Israeli drones and found them to be clean of US technology and on par with the drones China was offering to sell. While Washington did not like Israel's growing alliance with Russia, it grudgingly gave Jerusalem the green light to move ahead with the sale.

"Because it's a joint venture and Russia will have skin in the game, the Israelis thought they could later perhaps maneuver to have influence over Russian decision making in Iran and the larger region," Long recalled.

If a deal was happening, it might as well happen with Israel.

For nearly five years, Israel's drone deal seemed to do the trick. Iran kept pressuring Russia to deliver the S-300, but the Kremlin refused to supply the system. Reports that Iranians were training on the S-300 in Russia popped up now and again, rattling nerves in Jerusalem, but the system stayed grounded in Russia.

In the summer of 2015, though, everything changed. After nearly a year of negotiations, Western superpowers led by the United States—the P5+1—reached a historic deal with Iran to curb its nuclear program. With a deal in place, Russia again started making noise that there was no longer a reason to delay delivery. The contract, the state-run Russian Technologies Corporation said, was "back in force."

Israel began prepping for a new diplomatic fight. But then something unexpected happened. In September, Russia started bombing ISIS targets in Syria as part of an effort to salvage Bashar al-Assad's regime. It deployed squadrons of combat aircraft to Syria and dispatched navy warships and submarines in the Mediterranean. Israel hadn't seen the Russian military along its borders since the Soviets left Egypt in the 1970s. Concerned about a potential misunderstanding that could lead to a conflict, Israeli prime minister Netanyahu flew immediately to Moscow to set up, with President Vladimir Putin, a delicate coordination mechanism between the IDF and the Russian army.

In November, a Russian Sukhoi bomber was shot down by Turkey. Furious, Moscow threatened military action and suspended trade with Ankara. But then Putin did something else. He sent the S-400—an upgraded version of the S-300—not to Iran, the place whose acquisition of the system Israel had fought against for years, but to Syria, right along Israel's northern border, literally in the Israeli Air Force's backyard.

Like the S-300, the S-400 can track and intercept multiple targets from distances of hundreds of miles, but it also comes with an upgraded radar system more resistant to jamming as well as a variety of missiles that provide several layers of defense. With an extended range, the S-400 can shoot down planes over Tel Aviv all the way from Syrian territory.

For Israel, the news was shocking. "In our worst nightmares, we never thought this would happen," one senior IAF officer said.

Beyond the operational ramifications—the IAF needed to change some of its flight patterns—the Russian deployment in Syria showed Israel that arms sales are limited in their influence. In a region as complex as the Middle East, reality will always be stronger.

CONCLUSION

Armageddon and the Future of Weapons

It was planned to be Israel's 9/11, the kind of attack that would shock an entire nation. Years of wars, suicide bombings, rocket attacks and bloodshed were supposed to pale in comparison to what the Hamas commanders were planning for Israel in Gaza. This was supposed to be judgment day.

At about 4:30 a.m., the men started emerging from the ground. About a dozen came out of the hole dressed in army fatigues and armed to the teeth with AK47s, rocket-propelled grenades (RPGs), pistols, hand grenades and night-vision googles. Some had Go-Pro cameras attached to their helmets so they could film their operations. They came out of nowhere, smack in the middle of a cucumber field.

It was July 17, 2014, about a week into the latest Israeli-Hamas battle in the Gaza Strip. Dozens of rockets were being fired daily into Israeli towns, and air force jets were bombarding Gaza, hunting down Hamas commanders and hitting terror bases, rocket launchers, command posts and arms caches.

Israel had prepared for the tunnel attack but didn't know exactly where the exit would be. Intelligence pointed to an approximate location. The IDF was deployed heavily nearby while drones circled above.

In a black and white thermal video taken by one of the drones and later released by the IDF, the Hamas men are seen climbing out of a hole in the ground. They then spread out in the field until they realize they have been spotted. They rush back to the opening and, one by one, climb back underground. As the last man starts his descent, an Israeli missile strikes, destroying the tunnel entrance and killing several members of the cell inside.

The tunnel attack was attempted just hours before a cease-fire between Israel and Hamas was supposed to go into effect. For Israel, though, it was the straw that broke the camel's back. About 10 days before, Israel had bombed another massive tunnel that was discovered a bit further south, leading into a nearby kibbutz.

Israel knew that Hamas had been digging "attack tunnels" across the border and that it had created a secret commando unit called Nukhba—Arabic for "the selected ones"—whose top members were trained to move and fight through narrow tunnels, mostly on foot but even on small motorcycles.[1]

Hamas, the IDF claimed, had planned to infiltrate dozens of men through about 30 different tunnels, each strategically dug to end at the entrance to a different kibbutz or town. The men were supposed to enter private homes, dining halls and kindergartens and embark on a killing spree. One part of the team was supposed to grab a few Israeli men and women and force them back into the tunnel to Gaza, where they would be held as bargaining chips for future prisoner swaps. The goal: dozens dead and dozens more abducted. The images alone would have been a stab to Israel's national morale.

The existence of tunnels in Gaza was not a big secret. They had been part of the landscape since the 1980s, when the southern Gaza town of Rafah was split in two after Israel struck a peace deal with Egypt and returned the Sinai Peninsula. Then, though, the tunnels were mostly used to smuggle contraband across the border. One of the first known uses of tunnels by terrorists was in 1989, when Mahmoud al-Mabhouh, a Hamas terrorist, escaped Gaza via a tunnel after abducting and murdering two IDF soldiers. Mabhouh would go on to become one of Hamas's top operatives, responsible for weapons procurement. In 2010, he was assassinated by a believed-to-be Mossad hit team in his Dubai hotel room.

By the early 2000s, there were hundreds of operational tunnels along the Egyptian-Gaza border used to smuggle into Gaza anything that could fit—from arms to cigarettes, from explosives to plasma TVs and even cars.

Due to their location—right along the border with Egypt—residents of the Gaza town of Rafah became experts at digging tunnels, often using small children to construct the underground passageways and smuggle the contraband. The shafts were first dug inside homes near the border, and construction took anywhere from two weeks to two months, with costs sometimes reaching $100,000. While they were pricey to build, the revenue from a successful smuggling tunnel was potentially huge.

As the years went by, Hamas integrated larger and more sophisticated tunnels into its combat doctrine and used them to attack IDF forts inside the Gaza Strip. In 2004, for example, a powerful 1.5-ton bomb went off inside a tunnel under an IDF position near Rafah. Hamas commandos then stormed the post, killing five Israeli soldiers.

From a strategic perspective, the use of tunnels by Hamas made sense. While its rockets were doing an effective job of terrorizing Israeli civilians, the heavily guarded border made it dif-

An Israeli soldier stands over the entrance to a terror tunnel uncovered along the border with the Gaza Strip in 2014.   IDF

ficult to carry out large-scale strategic attacks that could shock the Jewish State. Infiltrating Israel above ground was virtually impossible thanks to a large fence bolstered by a sophisticated array of radars, observation posts and patrols. Underground passageways were a perfect way to enter Israel, carry out an attack and escape. The tunnels had immense potential.

The tunnel industry had undergone significant improvements. In October, a massive tunnel—never seen before by the IDF—was discovered on the Israeli side of the Gaza border near Ein Hashlosha, a pastoral kibbutz established in the 1950s by a group of new immigrants from Latin America. The discovery was pure luck— the combination of bits of intelligence and complaints from residents who were hearing strange noises.

The tunnel was 50 feet deep, ran for almost a mile and was six feet in height, tall enough for a person to stand up straight inside. Over 500 tons of concrete had been used to build the massive underground passageway, which one kibbutz resident said reminded him of the New York subway. Israeli intelligence estimated that the tunnel cost millions of dollars to construct and that Hamas had planned to use it to infiltrate the kibbutz and massacre as many residents as it could.

While Israel knew of the existence of the tunnels, nothing prepared it for the summer of 2014. After the attack on July 17, the Israeli Security Cabinet decided to send Israeli ground forces into Gaza to locate, map out and destroy the cross-border tunnels. There were two problems. Israeli intelligence didn't know where all the tunnels were located, and there was no detailed tactic for destroying them. While IDF troops understood their mission, they were all but clueless about how to complete it.

The troops were sent in for a limited and critical operation, but what they discovered shocked even the veterans. Hamas had built some 30 tunnels from Gaza into Israel. Most were already completed. Others were close. These tunnels were no longer the short and narrow ones Israel had discovered in the past. These had air ducts, cement walls, and communication lines and were large enough for people to stand up inside them. Some of the tunnels went on for miles, with several offshoots from the main line, meaning that even if you located one entrance, there were likely several others still out there. This meant that the IDF would need to go deep underground to map out a tunnel's full and intricate route.

That, in turn, according to a report in the *Wall Street Journal*, presented two key challenges. The first was how to locate and identify the tunnels, and the second was how to destroy them.

The IDF had a vague idea for the first challenge. Before the operation, using technology that sweeps cellular networks and signals, Israeli intelligence was able to identify the origin by tracking the diggers' cell phone signals. The signals, Israel discovered, would disappear when the diggers went underground and reconnect hours later from the same spot when they climbed back out.[2] That pattern usually meant that a tunnel entrance was nearby.

Since the early 2000s, the Defense Ministry had tested various systems to see if they could successfully detect a tunnel. A state comptroller report from 2007, though, blasted the ministry and the IDF for not doing enough. Either way, when the troops crossed into Gaza in 2014, they were working blindly, without any real technological help. At the beginning of the anti-tunnel operation, the defense minister predicted it would take two to three days. It ended up taking nearly three weeks, during which Hamas continued attacking via the tunnels.

Without a clear methodology to detect and destroy the tunnels, the IDF improvised, trying out different systems along the way. The first few tunnels discovered by Israel immediately had their shafts blown at the entrances the IDF troops exposed. But then the IDF realized that blowing one entrance didn't really do anything if the tunnel had several breakaway branches, as almost all of them did.

The soldiers then lowered robots inside and drove them along the tunnel routes. In other cases, they poured liquid explosives into the tunnels, and in rare instances, soldiers descended into the tunnels and lined the walls with dynamite and mines. The air force also assisted with the operation. After a tunnel route had been identified, jets would drop dozens of JDAMs—precision-guided bombs specially designed to penetrate the ground—along the tunnel line. The IDF called this "kinetic drilling."

But even as Israeli troops were operating in Gaza to locate and destroy the tunnels, Hamas still succeeded in carrying out a number of attacks. In one attack, a group of Hamas fighters exited a tunnel near the kibbutz of Nahal Oz. In a video Hamas later released and filmed on a helmet-mounted GoPro camera, the gunmen are seen exiting the tunnel, running through a field in broad daylight, infiltrating a nearby IDF border post and killing five soldiers. The infiltrators then run back to the tunnel and escape back to Gaza.

By the end of the operation, Israel had succeeded in destroying some 30 tunnels at a heavy price—dozens of soldiers were killed alongside hundreds of Palestinians. In addition, a military operation that should have taken just a few days ended up lasting 50, the longest conflict in Israeli history since the 1948 War of Independence.

The war was a wake-up call. Israel had been aware of the way Hamas used hospitals and schools to hide its weapons and command centers, but the tunnels it discovered represented a new level of fighting. These tunnels had the potential to enable a 9/11-scale attack.

What Israel encountered during the Gaza war of 2014 was a new type of warfare. The IDF thought it was prepared for war in Gaza. IDF soldiers were trained as experts in urban warfare; its tanks had been outfitted with new Trophy active protection systems to defend against anti-tank missiles; and the Iron Dome was proving to be more effective than expected in intercepting Katyushas and Kassams fired into Israel. But the tunnels made what the IDF had prepared for seem almost meaningless.

It was a classic case of "disruptive innovation," the term Harvard Business School professor Clayton Christenson coined to describe innovations (in this case terror tunnels) that make traditional competitors (in this case the IDF) irrelevant.

In the end, though, the IDF succeeded in its mission. It took longer than initially anticipated and cost more lives on both sides than anyone wanted, but the military ultimately found a way to locate and destroy the tunnels. It adapted to a changing battlefield and created a new knowledge base, now desired by militaries worldwide.

Israel might have been the first country to face terror tunnels along its border, but those same tunnels could one day pop up between America and Mexico, Turkey and Syria or along India's border with Pakistan. These are not theoretical threats.

Israel's experience during the Gaza war showed the IDF that as prepared as it might think it is for war, it can always be surprised.

It is not a new feeling for the country, and particularly the military. When the Yom Kippur War broke out in 1973, for example, IDF tank crews stationed along the Egyptian border started getting hit by a mysterious missile. At first, commanders thought the tanks were being hit by standard RPGs, fired by infantry soldiers. But when they pulled back a bit, supposedly out of range, the tanks were still getting nailed.

As the fighting continued, the IDF tankers finally understood that they were facing a new threat—a missile that was fired by a single soldier, was guided by a wire and could accurately hit targets from over a mile away. It was a new Soviet anti-tank missile called Sagger. To defeat the Sagger, the IDF had to improvise. Whenever a missile launch was detected, the tanks would alert one another and then together begin driving in random directions, kicking up clouds of sand to block the missile launcher's line of sight.

The tactic was successful and later adopted by NATO. While it proved effective, it didn't make the Israeli tanks immune; more than 180 Israeli tanks were knocked out in the first day of the war, half of them by Saggers.[3]

What Israel showed in 1973, in 2014 and throughout its history, though, is an ability to improvise in real time, under pressure and on the battlefield. As we have shown throughout this book, improvisation is a hallmark Israeli characteristic, one the country calls on often to confront different challenges and threats.

The lesson from these experiences is that as much as the IDF prepares, there will always be surprises. Israel thought it had stopped the delivery of the S-300 to Iran and then found an upgraded version of the system in Syria; it thought rocket fire from Gaza was not a strategic threat until thousands of rockets rained down on its cities. As the region continues to undergo one of the greatest upheavals in history, IDF commanders know that surprises are a sure thing. The best they can do is minimize their impact.

Preparing for uncertainty almost sounds like a paradox, but that is exactly what the IDF does on a daily basis. It cannot be sure exactly how the next war will look, but like all modern militaries, it tries its best to prepare for future wars and not those of the past.

The test bed for Israel's future conflict is located hundreds of feet underneath the Defense Ministry in Tel Aviv, in an underground command center known as the Bor, the Hebrew word for "pit."

The Bor is the IDF's nerve center, the place where all major operations are plotted and directed. It is accessed through two massive steel doors that are sealed shut in the event of a nonconventional attack. A big sign warns visitors to leave their cell phones outside. With Iran and Hezbollah actively eavesdropping on Israel, no chances are taken. The Bor has its own air-purification system and power source. Even if the buildings above ground are destroyed, the Bor can continue to function.

The stairs seem to go down for miles. On one floor there is a door with a sign reading "Northern Front—Syria." Down the hall are similar rooms for Gaza and Lebanon as well as for the front the IDF calls "Depth," places where troops might need to operate far from Israel's immediate borders. These rooms are where operations officers pore over maps, draft plans for future operations and decide which units and aircraft will be assigned to the different operations and battlefields.

A couple of more floors down is the IDF chief of staff's conference room. There, around the U-shaped table, the top military command convenes weekly to review pending operations and debate doctrines and tactics that can be applied in Israel's future conflicts. The walls are lined with photos documenting previous monumental meetings held in the room during Israel's various periods of crisis. The solemn and long faces of the generals in the pictures are a constant reminder of the decisions made in the room and the secrets its walls are meant to keep.

Along one side of the conference room is a ceiling-to-floor glass wall, separating the room from the IDF's main command center, better known as the War Room. The chief of staff has a seat in the middle of a long table lined with computers and phones of different colors according to their level of encryption. The generals sit facing a wall of screens, each showing a video feed from a different sensor—aircraft, naval vessels and satellites. This is where the chief of staff oversees live operations.

One summer day, a few years after the Second Lebanon War, a top IDF general stood in the conference room lecturing on Israel's future battlefield. The picture was gloomy. The upheaval in the Middle East, he explained, had its advantages but mostly presented Israel with new threats and challenges. On the one hand, the Syrian military, once Israel's primary adversary, no longer existed, meaning that there was no longer a real conventional military threat against the Jewish State. The days, he said, when Israel needed to worry about tank invasions and losing territory to its enemies appeared to be over.

On the other hand, the officer went on, Hezbollah and Hamas were no longer small-time terror groups. In future wars, he said, the IDF will face a "collage war" and will need to know how to cope with anti-tank missiles (conventional), abduction of soldiers (terror) and terrorists popping out of tunnels (guerilla) all at the same time. While combat aircraft are important for bombing strategic targets and projecting power throughout the wider region, they can't really do anything when 50 Hamas terrorists jump out of a tunnel and storm a nearby kibbutz dining hall.

To prepare for these "collage" type of conflicts, the IDF put an emphasis in recent years on three different areas: improving interoperability, increasing the use of standoff/robotic platforms and ensuring that Israel retains international legitimacy for its operations and actions.

"Interoperability" means the ability for military units from different disciplines to work together. At the most basic level, it means that pilots and infantry soldiers speak the same language so they can understand what the other is saying when soldiers on the ground are, for example, trying to direct a pilot to bomb a target. On a more advanced level it has direct technological implications. If, in the past, air force pilots could not see targets soldiers were fighting on the ground on the screens in their cockpits, today they can.

This increase in interoperability is thanks to officers like Colonel Hanan Iserovich, who until 2015 served as commander of

Mamram, the IDF's elite computer unit. A Hebrew acronym for "Center of Computing and Information Systems," Mamram has made a name for itself as a center of excellence, responsible for maintaining IDF networks and ensuring tactical connectivity. Its soldiers are scooped up by tech companies the moment they end their mandatory service.

One recent invention by the unit is a system the IDF calls Crystal Ball, which allows commanders to transfer coordinates of targets to digital maps by literally touching the target on a screen transmitting live video footage from a battlefield.

This allows officers sitting in command centers to select targets they see on video footage provided by any sensor—like a drone's camera—and have them turned immediately into digital coordinates, which are relayed to the digital mapping systems of all combat units operating within an area.

Crystal Ball and other systems like it help shorten the IDF's sensor-to-shooter cycle. Today, what used to take 20 minutes happens in a fraction of that time.

Iserovich knows how critical systems like these can be. In 2006, during the Second Lebanon War, he was commander of the Nachal Brigade's 50th Infantry Battalion and was deployed with his troops in southern Lebanon. One day, his radio crackled. On the other side of the line was an intelligence officer, stationed at Northern Command headquarters back in Israel, with a critical piece of intelligence.

"There is a Hezbollah anti-tank missile squad in a home nearby," the officer said. "Get ready."

Such intelligence was rare during the war. The IDF had decided to invade Lebanon to put an end to Hezbollah's open deployment along the border but did not have quality intelligence on the group's positions throughout southern Lebanon and its villages. This time, though, the intelligence was dead-on accurate. Using his night-vision goggles, Iserovich was able to identify the Hezbollah cell, located under a mile away. Time was of the essence. While Iserovich saw the cell, the Hezbollah fighters had not yet detected the IDF troops. Once that happened, it would be a full-fledged onslaught.

Iserovich raised an Israeli Air Force Apache attack helicopter flying nearby and asked the pilot to bomb the home where the Hezbollah cell was hiding. For 15 minutes, Iserovich spoke to the pilot, explaining again and again where the Hezbollah cell was located and where the IDF troops were located. The pilot wanted to make sure he wasn't going to accidentally target the home where Iserovich was hiding with his troops. It took over a dozen exchanges until the pilot was confident that he and Iserovich were talking about the same target.

Experiences like this led the IDF to the conclusion that it was severely lacking in interoperability. Now, with systems like Crystal Ball, all a commander like Iserovich needs to do is highlight the Hezbollah home on his electronic map, and then it is seen by all other units, including fighter jets and helicopters, sitting on the same network.

"The enemy today—whether Hezbollah or Hamas—has a low signature, is slippery and operates inside an urban setting," a senior IDF officer explained. "We need to know how to detect, identify and engage such targets quickly and accurately."

The second major change the IDF is undergoing is the integration of more autonomous systems—robots—into combat. By 2015, the majority of flights conducted by the Israeli Air Force were being done by drones. In the coming years, the number will continue to increase, and by 2030 the air force plans to have a fleet made up solely of drones and stealth fighter jets.

But drones are not just staying airborne. On the ground, the

IDF is using a variety of unmanned systems, from ball-shaped cameras that can be thrown into a room and roll around to provide a picture of what is happening inside, to unmanned ground vehicles that can patrol Israel's volatile borders with Gaza and Syria without putting soldiers at risk.

Another future option is to send the unmanned ground vehicles—known as UGVs—into enemy territory before troops. Risky missions that used to be carried out by elite reconnaissance teams can now be performed by a small unmanned dune buggy equipped with 360-degree cameras, loudspeakers and an automatic rifle. Storming homes without knowing what is happening inside is a thing of the past. Soon, robotic snakes will slither their way into enemy headquarters before soldiers storm the place. And then there are unmanned patrol ships like the Protector, developed by the Israeli company Rafael, that comes with a weapons

An IDF unmanned ground vehicle, called Guardium, patrols the border with the Gaza Strip. IDF

kit as well as a high-pressure hose for nonlethal missions. With a design based on that of a small speedboat, the ship has already undergone Israeli Navy trials off the Gaza coast.

In the not-too-distant future, robots—in the air and on the ground—will storm the battlefield at the front. If soldiers are ultimately deemed necessary, they will look starkly different from the soldiers of today. Instead of standard boots and combat vests, these soldiers will have robotic armor that provides protection but also allows troops to move faster and carry heavier loads without feeling the difference. They will have personal head display sets—increasing their situational awareness—and will be armed with assault rifles with bullets that can distinguish between enemy and friendly forces.

The tanks sent into battle will also be unmanned and will come with hybrid engines capable of operating nonstop for weeks on end. The drones in the air will operate on solar power and fuel cell technology installed on their wings. They won't have to land except for routine maintenance inspections.

Clusters of miniature nanosatellites, launched by the air force over battle zones, will be able to provide real-time imagery for ground forces as well as serve as "cell towers" for Israeli communication devices operating far from home. Precision rockets will strike predesignated targets, leaving the air force's stealth fighter jets available for longer-range, strategic missions.

If the Six Day War were fought again in the future, it is possible that Israel would not even need to send its jets to take out the Egyptian and Syrian air forces. It could alternatively keep all of its aircraft on the ground and use sophisticated and powerful cyber weapons instead.

This is the future of Israeli warfare.

✳ ✳ ✳

All of this, though, is meaningless if Israel's operations lack the international stamp of legitimacy. The state can develop, manufacture and even sell weapons around the world, but that won't mean much if the world refuses to support Israel's actions.

For a country like Israel, legitimacy is not trivial. Highly dependent on international—and particularly American—support, Israel has almost always sought approval, even if just tacit, before taking military action. This was the case in Israel's recent conflicts in Gaza and Lebanon but also when it came to Israeli deliberations regarding Iran's nuclear program and its ultimate decision to refrain from launching a military strike, strongly opposed by the rest of the world.

Wars, especially those against nonstate actors, do not end on the battlefield. They continue in the media, in the courtroom and around the United Nations Security Council table. Public support today means far more than it did a decade ago, and lack thereof can end a war before military objectives are met.

This is exactly what happened the morning of July 30, 2006, two weeks into Israel's war with Hezbollah. An explosion rocked the southern Lebanese town of Kfar Kana, caused by an Israeli Air Force bomb dropped the night before, which, due to some malfunction, had failed to explode immediately.

Initial reports spoke of more than 60 casualties—half of them children—and almost every international news network connected to the live feed Al-Jazeera was broadcasting from the rubble. While Israel claimed that the building's vicinity was being used as a launchpad for Katyusha rockets into Israel, it failed to present proof until over 12 hours later.

That day was a turning point in the war. US secretary of state Condoleezza Rice happened to be in Israel and used the bombing to get Israel to suspend all aerial activity over Lebanon for the next 48 hours, a steep price in the middle of a war. In the days that followed, international support for Israel's campaign against Hezbollah almost completely eroded.

It's possible that had Israel more quickly released evidence of what was happening in Kfar Kana, the support would have held together. Either way, Kfar Kana and other incidents since have forced Israel to adapt to a changing reality that affects not just its diplomatic corps, but also military officers who, before attacking targets, need to think about a camera that might be filming them at that exact moment—and about the international legal consequences the attack might have. Will that IDF officer be able to travel freely throughout the world, free of fear of arrest? Or will international warrants be issued for Israel's top officers?

When we arrived at the seaside apartment building just north of Tel Aviv, we weren't sure what to expect. Our meeting with Shimon Peres, Israel's former prime minister and president, was supposed to start in a few minutes. We didn't know, though, how much time the 93-year-old Peres would have for us and whether he would even remember the stories we were hoping to hear. We were in for a surprise.

Peres was the ultimate Israeli statesman. Scattered along the bookshelves that lined his apartment were artifacts from his 70 years in government—pictures with world leaders, awards from foreign countries and the Nobel Peace Prize he controversially shared with Yitzhak Rabin and Yasser Arafat.

If there was one Israeli who had seen it all, it was Peres. He was at Ben-Gurion's side throughout the War of Independence and was later the fledgling state's key arms buyer. It was Peres who persuaded Al Schwimmer to move to Israel and establish Israel Aerospace Industries, and it was again Peres who crafted Israel's strategic relationship with France, which culminated in the founding of the country's highly secretive nuclear program.

In government, he served in almost every ministry—transportation, defense, finance and foreign. He was twice Israel's prime minister and in his last job—between 2007 and 2014—the nation's president.

If there was someone, we thought, who could explain the secret to Israel's success in developing some of the world's most innovative weaponry, it was Peres.

Peres got his start as Ben-Gurion's assistant by pure luck. It was 1947, and the kibbutz where Peres grew up in the Jordan Valley decided to send the 24-year-old to Tel Aviv to volunteer with the Haganah. One Saturday night, the kibbutz secretariat held a vote, and the next morning, Peres was on his way to the "Red House" in Tel Aviv—the Haganah headquarters—with three liras his kibbutz friends had stuffed in his pocket.

But when he arrived, no one knew what to do with him. Wandering around the building, he bumped into an old friend. "Do you know what I'm supposed to do here?" Peres asked.

"No," his friend said. "We really don't have anything for you."

Peres began to panic slightly. Had he made the trip for nothing? What would he tell his kibbutz friends when he returned a day after they had sent him? As he stood contemplating what to do, he heard a loud voice from the staircase.

"Ah, you're here?" Peres turned around and saw David Ben-Gurion, Israel's new prime minister, standing before him. The two knew each other from Mapai, the left-leaning party they both belonged to at the time and which would several years later merge into what is known today as Israel's Labor Party.

"Yes," Peres responded. Ben-Gurion walked over, took a piece of paper out of his jacket and handed it to Peres. "This is a list of what we have in arms and weapons," Ben-Gurion said, rattling off the number of machine guns and bullets in Haganah's cache. "If we don't have weapons and are attacked, we will be destroyed. We need weapons. That is the most important mission. That is what you need to do."

Ben-Gurion's positive reception didn't help Peres. The other Haganah leaders continued to ignore him. Later in the day, his friend found him. He told Peres that while he didn't have a specific job for him, he could go sit in the office of Yaakov Dori, the Haganah chief of staff, who was homesick.

Peres went to Dori's office and took a seat behind his large wooden desk. Bored and not sure what to do, he sifted through a drawer and found inside two letters addressed to Ben-Gurion. He opened one and received the shock of his life.

In it, an IDF general explained to Ben-Gurion his decision regarding the prime minister's offer to serve as the new IDF chief of staff. The general wrote that while he was flattered by the offer, he had decided to turn it down after discovering that the Jewish State had only six million bullets in its arsenal.

"We will need 1 million bullets a day in a war and I am not willing to be chief of staff for just six days," the general wrote.

Peres hadn't realized the situation was so dire. His shock though didn't last long. There was work to do. Soon, he found himself at Ben-Gurion's side, running a network of agents around the world who were tasked with doing whatever was needed to obtain weapons and ammunition for the State of Israel. Peres would spend most of the next decade dedicated to that objective.

These experiences molded Peres's personality. While he worked tirelessly to obtain weapons, he also did a lot of dreaming, fantasizing about ways for Israel to overcome its dangerous military disadvantage. In those early years, for example, Israel didn't have tanks or aircraft or anyone willing to sell them. So

Peres came up with innovative ideas to circumvent the problems, like having Israel independently develop anti-tank missiles and anti-aircraft guns so that, at the very least, it could defend itself.

"If there is a wall, it is foolish to just knock your head up against it," he told us. "What you need to do is think of another way to get the job done. You need to be creative."

It wasn't always easy sailing. In the 1950s, for example, Israel started manufacturing its own rifles. But when the IDF took the rifle for trials, the bullets blew up inside the barrel. The soldiers fired again with a different rifle, but the bullets again exploded. The officers were mystified. They inspected the bullets and the rifles but couldn't determine the cause for the malfunction until one engineer decided to inspect the copper being used to make the bullets. It turned out that rats had infested the warehouse where the copper was stored and that rat urine weakened the copper, making the entire batch of bullets defective.

With Ben-Gurion at his side, Peres embarked on a journey to "scientificize" the country. When Peres came to the IDF and suggested buying supercomputers, he was thrown out of the room. When he proposed investing in the development of missiles, the generals again laughed. "We don't have bullets, and you are talking about missiles and computers," they said mockingly.

The obstacles didn't go away, but Peres was blessed with an unusual degree of persistence. When the finance minister, for example, told him that he wouldn't allocate "even one stinking penny" for the construction of a nuclear reactor, Peres succeeded in raising the millions of needed dollars off-budget. When Israel's universities refused to cooperate in the development of weapons, he found scientists elsewhere. Peres saw opportunity where others saw peril. He refused to give up or succumb to the name-calling and the insults he knew were being whispered behind his back.

"We were attacked from the beginning and didn't have a choice," he told us solemnly, recalling how he and Ben-Gurion sat together the night of November 29, 1947, when the UN voted on the partition plan. "Today they dance," Ben-Gurion said that night. "Tomorrow there will be war."

But somehow, Israel persevered and defeated its enemies. Peres said that the success was the result of a combination of high-quality people and an amazing degree of motivation.

"There is something in our DNA that makes us Jews never feel satisfied," he said. "Give a Jew something and he will add to it or fix it. Give the air force a plane and they will add to it and change it . . . We believe that anything is possible."

But, we asked the elder Israeli statesman, aren't you concerned about the erosion of Israel's qualitative military edge? In recent years, we said, the United States has announced plans to sell Saudi Arabia billions of dollars' worth of its most advanced military platforms. At the same time, Hezbollah and Hamas are becoming more military-like in their structures and are using weapons, like anti-tank missiles and drones, once reserved for conventional armies.

Peres thought for a moment and then laid out his vision for the future. "To retain Israel's qualitative edge," he said, "we need to invest in soldiers' brains, not just their muscles."

That was why Peres had recently proposed to the IDF chief of staff that he have every soldier attend university and get a BA degree before military service. That was why he had also recently recommended that the Education Ministry establish a program to teach two- and three-year-olds a second language in nursery school.

Nothing is impossible, Peres said, but he added that there is also nothing that happens on its own. If you want something to happen, you have to push forward, sometimes all by yourself.

Peres, we learned that day, was a dreamer. The man who built

Israel's military and purported nuclear capabilities talked about science, technology, robots and peace. The future of weapons and warfare, he told us, is in space, where the "potential is endless," and in understanding how to use less energy. "How to do more with less," he said, speaking enthusiastically about nanotechnology and the opportunities it provides Israel in developing smaller, smarter and more reliable sensors and weapons.

Years of wars, he said as we neared the end of our conversation, have made Israelis pessimistic and suspicious. Technology, he insisted, can change that. "The future does not happen like a Swiss watch," he said. "But we can change people's character. We can use technology to make people better, to live better lives and to be more hopeful about the future."

Whether Peres was right or wrong doesn't make a difference. We were just impressed that at 93, the former president still had the chutzpah to think 50 years ahead.

Working on this book, we have had a feeling that these are historic times. The combination of being journalists in Israel covering daily aspects of the Israel-Arab conflict and watching the development and introduction of new weapons into the region has given the impression of a complex but mostly fascinating era.

Israel's weapons are revolutionizing the modern battlefield. Their impact on the way wars are being fought is radiating far beyond Israel's borders to the wider Middle East, Europe, Africa and Asia. The story of those weapons has captivated the world with its unique combination of war-earned military experience and persistent innovation.

We wrote this book while skipping between countries—in Asia, Europe and of course Israel and the United States. In all of the different places we visited, the average person is vaguely familiar with Israel but rarely with its technological prowess and advanced weaponry.

We finished the book convinced that barely anybody in Israel wants war but that, as a country that has fought one every decade since its establishment and still has enemies along its borders who call for its destruction, Israel will always be prepared. We do not pretend to know the future, but we have no doubt that Israel's weapons will help shape that story.

ACKNOWLEDGMENTS

This book would not have been possible without the hundreds of people who agreed to speak to us and share with us their insights. Many of them spoke about the roles they have played in Israel's military and defense establishment. Others were diplomats, fighting wars in the global corridors of power. Some asked for anonymity, a request we have respected.

Yaakov owes a huge debt of gratitude to Ann Marie Lipinski and the rest of the amazing staff at Harvard University's Nieman Foundation for Journalism, where the idea for this book started to take form. The writing workshops he took there with the fabulous duo—Anne Bernays and Paige Williams—are evident in the book before you. He also thanks the *Jerusalem Post*, where he currently serves as editor-in-chief and which has provided him with a platform to tell Israel's story to the world for the greater part of the last 15 years.

Amir feels a debt of gratitude to *Walla*, his home for the past six years and Israel's leading news website. Amir came to *Walla* after 12 years at *Maariv* and found a news organization where he could take his storytelling to new heights in partnership with a first-class, cutting-edge and creative team.

Amir also thanks Bar-Ilan University, where he is writing his dissertation under Professor Shlomo Shapira, one of Israel's leading experts in the world of intelligence and terror.

A special thanks goes to our editor, Elisabeth Dyssegaard, who was excited about this book from the beginning, as well as to our dedicated agents, Peter and Amy Bernstein, who helped sharpen our idea and accompanied us throughout the process. Thank you all.

Last but not least, we owe a profound debt of gratitude to our families. Chaya (Yaakov's wife) and Fani (Amir's wife) gave us the support, space and time we needed—while taking care of our seven children (Yaakov—Atara, Miki, Rayli and Eli and Amir—Ron, Yahli and Tamari)—to get the job done. Chaya and Fani have always pushed us to be the best we can be. Without them, this book would not have been possible.

NOTES

INTRODUCTION

1. Gili Cohen, "Israeli Defense Exports in 2014: $5.6 Billion" [Hebrew], *Haaretz*, May 21, 2015, http://www.haaretz.co.il/news/politics/1.2642295.
2. Marcus Becker, "Factory and Lab: Israel's War Business," *Der Spiegel*, August 27, 2014, http://www.spiegel.de/international/world/defense-industry-the-business-of-war-in-israel-a-988245.html.
3. Fareed Zakaria, "Israel Dominates the Middle East," *Washington Post*, November 21, 2012, https://www.washingtonpost.com/opinions/fareed-zakaria-israel-dominates-the-middle-east/2012/11/21/d310dc7c-3428-11e2-bfd5-e202b6d7b501_story.html.
4. Reuven Gal, *A Portrait of the Israeli Soldier* (Westport, CT: Greenwood Press, 1986), 10.
5. Arthur Herman, "How Israel's Defense Industry Can Help Save America," *Commentary*, December 1, 2011.
6. Becker, "Factory and Lab."
7. Ann Scott Tyson, "Youths in Rural US Are Drawn to Military," *Washington Post*, November 4, 2005.

8. Christopher Rhoads, "How an Elite Military School Feeds Israel's Tech Industry," *Wall Street Journal*, July 6, 2007, A1.
9. Ben Caspit, "Talpiot Industrial Zone," *Maariv*, March 29, 2010, 10.

CHAPTER 1

1. Golda Meir, *My Life* (New York: Dell, 1975), 213, 222, 224.
2. Yuval Steinitz, "The Growing Threat to Israel's Qualitative Military Edge," Jerusalem Center for Public Affairs, *Jerusalem Issue Brief*, Vol. 3, No. 10, December 11, 2003.
3. Ignacia Klich, "The First Argentine-Israeli Trade Accord: Political and Economic Considerations," *Canadian Journal of Latin American and Caribbean Studies*, Vol. 20, 1995; and Shimon Peres and David Landau, *Ben-Gurion: A Political Life* (New York: Schocken Books, 2011), 16.
4. Michael Bar Zohar, *Shimon Peres: The Biography* (New York: Random House, 2007), 81.
5. Ibid., 77.
6. Ibid., 106.
7. Avner Cohen, *Israel and the Bomb* (New York: Columbia University Press, 1998), 53.

CHAPTER 2

1. "The Dronefather," *The Economist*, December 1, 2012, http://www.economist.com/news/technology-quarterly/21567205-abe-karem-created-robotic-plane-transformed-way-modern-warfare.
2. Richard Whittle, "The Man Who Invented the Predator," *Air and Space Magazine*, April 2013.
3. "Military UAVs: Up in the Sky, an Unblinking Eye," *Newsweek*, May 31, 2008, http://www.newsweek.com/military-uavs-sky-unblinking-eye-89463; and Phil Patton, "Robots with the Right Stuff," *Wired*, March 1, 1996, http://archive.wired.com/wired/archive/4.03/robots_pr.html.
4. One of the authors interviewed Karem in the summer of 2013.
5. Frank Strickland, "The Early Evolution of the Predator Drone," *Studies in Intelligence*, Vol. 57, No. 1, March 2013.
6. George Arnett, "The Numbers Behind the Worldwide Trade in Drones," *Guardian*, March 16, 2015.

7. See Anshel Pfeffer, "WikiLeaks: IDF Uses Drones to Assassinate Gaza Militants," *Haaretz*, September 2, 2011, http://www.haaretz .com/news/diplomacy-defense/wikileaks-idf-uses-drones-to -assassinate-gaza-militants-1.382269.
8. Nick Meo, "How Israel Killed Ahmed Jabari, Its Toughest Enemy in Gaza," *Daily Telegraph*, November 17, 2012, http://www.telegraph .co.uk/news/worldnews/middleeast/israel/9685598/How-Israel -killed-Ahmed-Jabari-its-toughest-enemy-in-Gaza.html.
9. The following story about Israel's reported strike in Sudan is based on foreign media reports, including those in *Time Magazine* and the *Sunday Times*, as well as the authors' understanding of how an operation like this potentially would have been conducted.
10. "How Israel Foiled an Arms Convoy Bound for Hamas," *Time Magazine Online*, March 30, 2009, and Uzi Mahnaimi, "Israeli Drones Destroy Rocket-Smuggling Convoys in Sudan," *Sunday Times*, March 29, 2009, http://www.thesundaytimes.co.uk/sto/news/world _news/article158293.ece.
11. Mahnaimi, "Israel Drones Destroy Rocket-Smuggling Convoys in Sudan."
12. WikiLeaks, https://wikileaks.org/plusd/cables/09KHARTOUM249 _a.html.
13. See WikiLeaks diplomatic cable 09KHARTOUM249 created February 24, 2009, https://wikileaks.org/cable/2009/02/09KHARTOUM249 .html.
14. As noted in endnote 9, the section on Sudan was based on foreign media reports, including those in *Time Magazine* and the *Sunday Times*, as well as the authors' understanding of how an operation like this potentially would have been conducted.

CHAPTER 3

1. Amnon Barzilai, "Turret Exposed" [Hebrew], *Globes*, July 29, 2006, http://www.globes.co.il/news/article.aspx?did=1000137025.
2. Josh Mitnick, "Mighty Merkavas Fail in War Gone Awry: 'Boom, Flames and Smoke,'" *Observer*, August 21, 2006, http://observer .com/2006/08/mighty-merkavas-fail-in-war-gone-awry-boom -flames-and-smoke-2/.

CHAPTER 4

1. Deganit Paikowsky, "From the Shavit-2 to Ofeq-1, a History of the Israeli Space Effort," *Quest*, Vol. 18, November 2, 2011.
2. See Paikowsky, "From the Shavit-2 to Ofeq-1"; and Y. Rabin, Diary, Tel Aviv, *Maariv*, 1979 [Hebrew], Vol. 2, 497–498.
3. Bob Woodward, "CIA Sought 3rd Country Contra Aid," *Washington Post*, May 19, 1984, A13.
4. E. L. Zorn, "Israel's Quest for Satellite Intelligence," https://www.cia.gov/library/center-for-the-study-of-intelligence/kent-csi/vol44no5/pdf/v44i5a04p.pdf.
5. Bernard Gwertzman, "Israel Asks US for Gift of Jets, Citing Saudi Sale," *New York Times*, April 4, 1981, 2.
6. Amnon Barzilai, "Here We Build a Force Multiplier" [Hebrew], *Haaretz*, September 25, 2001, http://www.haaretz.co.il/misc/1.736130.
7. Glenn Frenkel, "Israel Puts Its First Satellite into Orbit," *Washington Post*, September 20, 1988, A16.
8. Lawrence Wright, *Thirteen Days in September* (New York: Knopf, 2014), 35.
9. Moshe Nissim, "Leadership and Daring in the Destruction of the Israeli Reactor," *Israel's Strike Against the Iraqi Nuclear Reactor 7 June, 1981* (Jerusalem: Menachem Begin Heritage Center, 2003), 31.
10. Shlomo Nakdimon, "Begin's Legacy: 'Yehiel, It Ends Today,'" *Haaretz*, February 22, 2010.
11. Barzilai, "Here We Build a Force Multipier." See also "Meeting Minutes Regarding Israel-South Africa Agreement," Woodrow Wilson Center Digital Archive, http://digitalarchive.wilsoncenter.org/document/114148.
12. Amnon Barzilai, "Somewhere Beyond the Horizon," *Haaretz*, September 26, 2001.
13. Paikowsky, "From the Shavit-2 to Ofeq-1."
14. Uzi Eilam, *Eilam's Arc* (Eastbourne, UK: Sussex Academic Press, 2011), 232–237.
15. Ibid.
16. "We Operate on the Border of Imagination," Mako News Website [Hebrew], March 27, 2014, http://www.mako.co.il/pzm-units/intelligence/Article-0b71c430eb30541006.htm.

CHAPTER 5

1. Amnon Barzilai, "How to Build a Wall" [Hebrew], *Haaretz*, November 13, 2002, http://www.haaretz.co.il/misc/1.839763.
2. Lior Avni, "A Decade under Fire: 10 Years to the First Kassam" (Hebrew), *NRG*, April 14, 2011, http://www.nrg.co.il/online/1/ART2/232/334.html.
3. David Horovitz, "Only a Drill?" *Jerusalem Post*, May 25, 2010.
4. Anshel Pfeffer, "Behind the Scenes of Iron Dome" [Hebrew], *Haaretz*, November 23, 2012, http://www.haaretz.co.il/news/politics/1.1871793.

CHAPTER 6

1. See "Report of the Special Rapporteur on Extrajudicial, Summary or Arbitrary Executions," May, 2010, http://www2.ohchr.org/english/bodies/hrcouncil/docs/14session/A.HRC.14.24.Add6.pdf.
2. Laura Blumenfeld, "In Israel, a Divisive Struggle over Targeted Killing," *Washington Post*, August 27, 2006, https://www.washingtonpost.com/archive/politics/2006/08/27/in-israel-a-divisive-struggle-over-targeted-killing/2e6d9107-6a81-4500-a7e4-001b4fc853c9/.
3. Steven R. David, "Fatal Choices: Israel's Policy of Targeted Killing," *Mideast Security and Policy Studies*, No. 51 (Ramat, Israel: The Begin-Sadat Center for Strategic Studies, Bar-Ilan University, 2002).
4. Yaakov Katz, "Analysis: Lies, Leaks, Death Tolls & Statistics," *Jerusalem Post*, October 29, 2010, 1.

CHAPTER 7

1. David Sanger, *Confront and Conceal: Obama's Secret Wars and Surprising Use of American Power* (New York: Crown, 2012), 188.
2. Ibid., 195.
3. Peter Beaumont, "Stuxnet Worm Heralds New Era of Global Cyberwar," *Guardian*, September 30, 2010.
4. Ibid.

5. Ralph Langner in interview with one of the authors by phone from Germany.

6. John Markoff, "In a Computer Worm, a Possible Biblical Clue," *New York Times*, September 29, 2010.

7. Ellen Nakashima, "U.S., Israel Developed Flame Computer Virus to Slow Iranian Nuclear Efforts, Officials Say," *Washington Post*, June 19, 2012.

8. James Clapper, testimony before the Senate Select Committee on Intelligence, January 31, 2012.

9. Barbara Opall-Rome, "Israeli Cyber Exports Double in a Year," *Defense News*, June 3, 2015, http://www.defensenews.com/story /defense/policy-budget/cyber/2015/06/03/israel-cyber-exports -double/28407687/.

10. The following story about Israel's reported strike in Syria is based on foreign media reports, including those in *Der Spiegel* and the *New Yorker*, as well as the authors' understanding of how an operation like this potentially would have been conducted.

11. Sharon Weinberger, "How Israel Spoofed Syria's Air Defense System," *Wired*, October 4, 2007, http://www.wired.com/2007/10/how -israel-spoo/.

12. Ibid.

13. Eric Follath, "The Story of Operation Orchard," *Der Spiegel*, November 2, 2009, http://www.spiegel.de/international/world/the-story-of -operation-orchard-how-israel-destroyed-syria-s-al-kibar-nuclear -reactor-a-658663-2.html.

14. Ibid.

15. David Makovsky, "The Silent Strike," *New Yorker*, September 17, 2012.

16. Ibid.

17. Ibid.

18. Ibid.

CHAPTER 8

1. Yoram Evron, "Sino-Israel Relations: Opportunities and Challenges," *INSS Strategic Assessment*, Vol. 10, No. 2, August 2007, http://www.inss.org.il/index.aspx?id=4538&articleid=1479.

2. "Weizman Initiated Eisenberg's Involvement in Chinese Arms Sales 20 Years Ago," *Globes* [Hebrew], February 4, 1999, http://www.globes.co.il/news/article.aspx?did=82076.
3. Thomas Friedman, "Israel and China Quietly Form Trade Bonds," *New York Times*, July 22, 1985.
4. Amon Barzilai, "The Phalcons Didn't Fly," *Haaretz*, December 26, 2001.
5. A. M. Rosenthal, "On My Mind; The Deadly Cargo," *New York Times*, October 22, 1999.
6. Sharon Samber, "Congress Urged Not to Link Israel Aid to China Arms," Jewish Telegraphic Agency, June 13, 2000.
7. Sadanand Dhume, "Revealed: The India-Israel Axis," *Wall Street Journal*, July 23, 2014.
8. The following section on Israel's ties with Singapore is largely based on the autobiography of Singapore's founding father, Lee Kuan Yew: *From Third World to First: The Singapore Story—1965–2000* (New York: Harper, 2000).
9. Amnon Barzilai, "Israeli Officers Reveal: This Is How We Founded the Singapore Military" (Hebrew), *Haaretz*, July 15, 2005.
10. Lee Kuan Yew, *From Third World to First*, 15.
11. See Barzilai, "Israeli Officers Reveal."
12. Ibid.
13. Duncan Lennox, ed., "AGM-142 Popeye ½ (Have Nap/Have Lite/Raptor/Crystal Maze) (Israel), Offensive Weapons," *Jane's Strategic Weapon Systems*, Issue 50 (Surrey: Jane's Information Group, January 2009), 78–80. See also Nuclear Threat Initiative's Israel Section: http://www.nti.org/country-profiles/israel/delivery-systems/.
14. Nicholas Clayton, "How Russia and Georgia's 'Little War' Started a Drone Arms Race," *Global Post*, October 23, 2012.
15. WikiLeaks cable 09TELAVIV2757_a, https://wikileaks.org/plusd/cables/09TELAVIV2757_a.html.
16. WikiLeaks cable 08Moscow2785, https://wikileaks.org/plusd/cables/08MOSCOW2785_a.html.
17. WikiLeaks cable 09MOSCOW2800_a, https://search.wikileaks.org/plusd/cables/09MOSCOW2800_a.html.
18. WikiLeaks cable 09MOSCOW1111_a, https://wikileaks.org/plusd/cables/09MOSCOW1111_a.html.

19. WikiLeaks cable 09ABUDHABI192_a, https://wikileaks.org/plusd /cables/09ABUDHABI192_a.html.
20. WikiLeaks cable 09TELAVIV1340_a, https://search.wikileaks.org /plusd/cables/09TELAVIV1340_a.html.
21. WikiLeaks cable 09TELAVIV1688_a, https://search.wikileaks.org /plusd/cables/09TELAVIV1688_a.html.
22. Interview with Mary Beth Long in November 2015.
23. See WikiLeaks cable 09TELAVIV2757_a.

CONCLUSION

1. Adam Ciralsky, "Did Israel Avert a Hamas Massacre?" *Vanity Fair*, October 21, 2014, http://www.vanityfair.com/news/politics/2014 /10/gaza-tunnel-plot-israeli-intelligence.
2. Asa Fitch, "Early Failure to Detect Gaza Tunnel Network Triggers Recriminations in Israel," *Wall Street Journal*, August 10, 2014.
3. Saul Singer and Dan Senor, *Start Up Nation* (New York: Twelve Books, 2009), 42.

INDEX

Page numbers in italics refer to figures.